Clay Science and Technology

Edited by Gustavo Morari Do Nascimento

Published in London, United Kingdom

IntechOpen

Supporting open minds since 2005

Clay Science and Technology
http://dx.doi.org/10.5772/intechopen.83317
Edited by Gustavo Morari Do Nascimento

Contributors
Tanushree Choudhury, Celia Marcos, Ricardo I. Jeldres, Matías Jeldres, Herbet Alves de Oliveira, Cochiran Pereira dos Santos, Nandakumar Selvasudha, Unnikrishnan-Meenakshi Dhanalekshmi, Sekar Krishnaraj, Yogeeswarakannan Harish Sundar, Nagarajan Sri Sridurga Devi, Irisappan Sarathchandiran, Yıldırım İsmail İsmail Tosun, Gustavo Morari Do Nascimento

First published in London, United Kingdom, 2021 by IntechOpen
IntechOpen is the global imprint of INTECHOPEN LIMITED, registered in England and Wales, registration number: 11086078, 5 Princes Gate Court, London, SW7 2QJ, United Kingdom
Printed in Croatia

British Library Cataloguing-in-Publication Data
A catalogue record for this book is available from the British Library

Additional hard and PDF copies can be obtained from orders@intechopen.com

Clay Science and Technology
Edited by Gustavo Morari Do Nascimento
p. cm.
Print ISBN 978-1-83968-209-4
Online ISBN 978-1-83968-210-0
eBook (PDF) ISBN 978-1-83968-211-7

For EU product safety concerns:
IN TECH d.o.o., Prolaz Marije Krucifikse Kozulić 3, 51000 Rijeka, Croatia, info@intechopen.com or visit our website at intechopen.com.

Meet the editor

Dr. Gustavo Morari do Nascimento is a professor at the Federal University of ABC, Brazil. He has strong experience in many fields related to the spectroscopic and microscopic characterization of nanomaterials. He obtained a doctoral degree from the University of São Paulo (USP) in 2004 with a thesis on spectroscopic characterization of nanocomposites formed by conducting polymers and clays. He spent 2007–2008 working on a post-doctoral degree at Massachusetts Institute of Technology (MIT) in the resonance Raman study of double-walled carbon nanotubes doped with halogens under the guidance of the legendary Mildred S. Dresselhaus. Back in Brazil, Dr. Morari do Nascimento spent three years (2009–2011) at the Federal University of Minas Gerais (UFMG) working with the synthesis of nanostructured carbon modified with molecular magnets. Currently, his research focus is on using different spectroscopic techniques for molecular characterization of chemically modified, carbon-nanostructured materials and polymer nanocomposites. He employs Resonance Raman and surface enhanced Raman spectroscopy (SERS) coupled to microscopy techniques added to X-ray absorption techniques at the National Synchrotron Light Laboratory in his investigation.

Contents

Preface

This book presents the state-of-the-art results of synthesis, characterization, modification, and technological applications of clays, clay minerals, and materials based on clay minerals, such as polymer–clay nanocomposites and clay hybrids. The main goal of this work is to contribute to the rationalization of some important results obtained in the broad area of clays and clay materials characterization. Moreover, this book provides a comprehensive account of polymer and biopolymer–clay nanocomposites, use of clay as an adsorption material for industrial pollutants, the ceramics industry, and the physical–chemical aspects of aqueous dispersions of clay and clay minerals. This book is beneficial for students, teachers, and researchers who are interested in expanding their knowledge about the use of clays in a diverse range of fields, including nanotechnology, biotechnology, environmental science, industrial remediation, pharmaceuticals, and so on.

The overall idea of the book is to provide the reader with a comprehensive, up-to-date, and evidence-based overview of the current literature in clay science and technology. This book discusses the main and new aspects of clay science and technology and is divided into two sections: "Properties and Materials" and "Applications."

Properties and Materials

Chapter 1 discusses the use of molecular and vibrational spectroscopies in the characterization of polymer–clay nanocomposites. It presents and reviews the main results collected in more than a decade. The screening of the electronic and vibrational structures of polymer–clay nanocomposites through resonance Raman and X-ray absorption spectroscopies has been decisive in determining their structure and in the study of the interactions between clays and intercalated polymers in a myriad of synthetic conditions.

Chapter 2 reviews clay hybrid materials for the most conventional and widely used classes of polymers. It also discusses the general reasons for using clay in hybrid materials. The green composite formed by clay polymer mixing has many improved properties such as high glass transition temperature (Tg), high flame resistance, high tensile strength, and improved barrier properties, all of which may be applied in the textile and automobile industries as well as in the fields of environmental science and polymer engineering.

Chapter 3 discusses the structural modifications of vermiculite clays after some physical and chemical treatments. The vermiculite transformation by structural water loss occurs with increasing temperature, vacuum, irradiation with microwaves or ultraviolet, and with either alcohol or acidic treatment. On the contrary, the transformation by water gain occurs in vermiculites treated with hydrogen peroxide and in those subjected to ionic metal exchange.

Chapter 4 focuses on the treatment of contaminated slurries using different materials. The objective of this study is to determine the effects of seepage flow to the surface and groundwater from industrial discharge. In this study, important investigations are made on the production of composite granules with Şırnak shale and zeolite feed in order to activate microwave heavy metal sorption in the water compost system.

Applications

Chapter 5 focuses on the use of calcareous clays in the ceramic industry. They must be used with care in ceramics because the carbonate particles do not have the main reductions, which can make recovery difficult during firing. In addition, undesirable damage is possible when in contact with water. This chapter discusses these potential issues, which may occur during the burning of clays with increasing percentages of carbonates.

Chapter 6 broadly discusses the use of clays in the field of pharmaceuticals. The understanding of surface chemistry and particle size distribution of clay minerals has led the pharmaceutical field in many directions and future perspectives. Their unique structure, which helps clays to absorb material onto their layered sheets, has opened up a wide variety of applications in drug delivery. Their ability to control and alter drug release profiles can be exploited in many ways to design effective drug delivery systems. Further advancements in nanotechnology have helped to synthesize and modify clay minerals to enhance their physiochemical properties and their usage as excipients.

Chapter 7 examines the difficulties in the mining industry due to water scarcity and the increased necessity of more sustainable processes. The use of seawater in this industry could be a potential turning point; however, the impact on clay properties under saline conditions must be better studied and analyzed. The rheological data presented by the authors are very interesting and elucidative.

Even though the knowledge regarding the real impact of materials derived from clay on the environment, industrial processes, and human health is still limited, I hope that this book is a useful tool for understanding the broad and complex field of clay science and technology. I would like to thank all the authors for their contributions. I am also grateful to the editorial staff at IntechOpen for their support throughout the process of publishing this book.

Dr. Gustavo Morari do Nascimento
Professor,
Centre for Natural Sciences and Humanities,
Federal University of ABC,
Santo André, Brazil

Section 1

Properties and Materials

Chapter 1

Two Spectroscopies as Main Source for Investigation of Polymer-Clay Materials

Gustavo Morari do Nascimento

Abstract

In the recent years the synthesis and characterization of nanomaterials has been one of the most efficacious way to produce new materials with improved or completely new properties. The polymer-clay nanocomposites are one of the most interesting nanomaterials with the possibility to create a myriad of new materials with many applications. Lamellar materials are classified as two-dimensional (2D), because there are formed by platelets piled up in one crystallographic direction, as the graphite and clays. The synthesis of controlled dimensional nanostructures as well as the characterization of the intrinsic and potentially peculiar properties of these nanostructures are central themes in nanoscience. The study of different nanostructures has great potential to test and understand fundamental concepts about the role of particle dimensionality on their physicochemical properties. Among the various materials studied in the literature, undoubtedly, polymer-clay materials, especially conducting polymers with smectite clays, such as montmorillonites (MMT) are of particular note. Our group have paid many efforts in the characterization of nanomaterials by using powerful spectroscopic techniques to study both the guest and host in case of inclusion compounds, nanofibers, carbon allotropes or many phases present in polymer-clay nanocomposites. There are two central questions that it was possible to address in this study: (i) the molecular structure of the polymer is drastically changed inside the interlayer cavity of clay and (ii) by using the appropriate synthetic or heating route is possible to change the molecular structure of the confined polymer. In the follow lines, it is briefly told the main aspects of resonance Raman and X-ray absorption spectroscopies in the study of polymer-clay nanocomposites.

Keywords: clay, nanocomposites, raman, XANES

1. Introduction

1.1 Clay science

Probably the clay is one of the most ancient and important material used and transformed by the humankind in order to produce a myriad of objects with many purposes. In fact, the historical impact of clay can be weighted by their intense use in many passages of one of the most influential book, the biblical text, as a synonym of a material that can be forged and transformed, as follows:

"Then the Lord God formed the man of dust from the ground and breathed into his nostrils the breath of life, and the man became a living creature." Genesis 2: 7 [1].

"But now, O Lord, you are our Father; we are the clay, and you are our potter; we are all the work of your hand." Isaiah 64: 8 [1].

In fact, farmers to produce plants explore the mechanical and chemical environment of clays, ceramists and artists continuously use clays to create extraordinary objects. To the editor, give softness to the paper surface in high quality prints. In medical area may be a relief for diarrhea and so on. In fact, there is no uniform nomenclature for clay and clay materials [2–4]. Clay material is "...a naturally occurring material composed primarily of fine-grained minerals, which is generally plastic at appropriate water contents and will harden with dried or fired". Naturally, this definition is elastic, because in geology science is considered clay the particles with size dimension of less than <4 μm, while in colloid science the value <1 μm is more acceptable [5]. The term clay mineral signifies a class of "...phyllosilicate minerals and minerals which impart plasticity to clay and which harden upon drying or firing" [6]. Since the origin of the mineral is not part of the definition, clay mineral (unlike clay) may be synthetic.

Hence, clay minerals have layers ordered in nanoscale and many different components can be present, as consequence, only by using advanced spectroscopic techniques it is possible to study their structures in detail. X-ray diffraction techniques are applied to determine the crystalline phases and basal distances d_{001}. The d_{001} is an important parameter to follow in the intercalation process. Clays layers have structures builded from tetrahedral sheets in which a silicon atom is surrounded by four oxygen atoms and octahedral sheets in which a metal like aluminum or magnesium is surrounded by eight oxygen atoms [7–10]. The tetrahedral (T) and octahedral (O) sheets are bonded by the oxygen atoms. Unshared oxygen atoms are present in hydroxyl form (see **Figure 1**). Two main arrangements of T and O layers can be observed in the structures of clays. One tetrahedral fused to one octahedral (1:1) is called as kaolin group with the general composition of $Al_2Si_2O_5(OH)_5$ and the layer thickness of ~0.7 nm. The crystal lattice consisted of one octahedral sheet

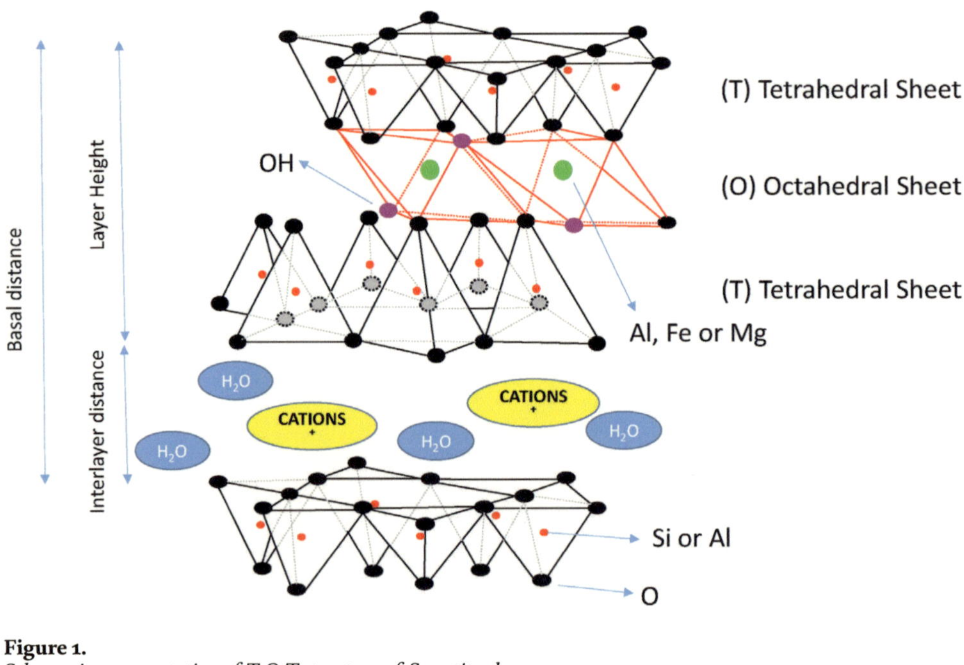

Figure 1.
Schematic representation of T:O:T structure of Smectite clay group.

sandwiched between two tetrahedral sheets (2:1) with the total thickness of 0.94 nm is well known as phyllosilicates. The 2:1 phyllosilicates those are electrostatically neutral with no inter layer ion and no expansion in water are known as pyrophyllite. However, when silicon in T sheets is substituted by aluminum, the 2:1 structure is called mica. The negative charge induced by this change is balanced by the introduction of potassium cations between the layers. Potassium cation has similar size of the hole created by Si/Al in tetrahedral sheets. Consequently, the 2:1 layers are held together strongly and the swelling or exfoliation of layers is not possible. The aluminum cations in the O layers can be partially substituted by divalent magnesium or iron cations in neutral pyrophyllite and as result the smectite clay group is formed, whose structure consists of a central sheet containing groups $MO_4(OH)_2$ of octahedral symmetry associated with two tetrahedral sheets (MO_4) producing layers designated T:O:T (see **Figure 1**). The O sites are occupied by ions of aluminum, iron and/or magnesium, while the centers accommodate tetrahedrons of silicon and aluminum ions.

In the last decades, the synthesis and characterization of nanomaterials and nanocomposites with improved or new properties has made the possibility of producing intelligent materials real [11]. One group of interesting nanomaterials with the possibility to create a myriad of new materials with many applications is the polymer-clay nanocomposites. Lamellar materials are classified as two-dimensional (2D), because there are formed by platelets piled up in one crystallographic direction, as the graphite and clays [12, 13]. The synthesis of controlled dimensional nanostructures as well as the characterization of the intrinsic and potentially peculiar properties of these nanostructures are central themes in nanoscience [14]. The study of different nanostructures has great potential to test and understand fundamental concepts about the role of particle dimensionality on their physicochemical properties. Among the various materials studied in the literature, undoubtedly, polymer-clay materials, especially conducting polymers with smectite clays, such as montmorillonites (MMT) are of particular note [15–25].

Our group have paid many efforts in the characterization of nanomaterials by using powerful spectroscopic techniques to study both the guest and host in case of inclusion compounds, [26] nanofibers, [27–29] carbon allotropes [30–38] or many phases present in polymer-clay nanocomposites [15–25]. In this brief chapter, we give an overview of some contribution of our studies of polymer-clay nanocomposites by using resonance Raman and X-ray absorption spectroscopies as main techniques of investigation. There are two central questions that was possible to address in our studies: (i) the molecular structure of the polymer is drastically changed inside the interlayer cavity of clay and (ii) by using the appropriate synthetic or heating route is possible to change the molecular structure of the confined polymer.

1.2 Techniques

1.2.1 Resonance Raman spectroscopy

Since the foundation of modern basis of physical sciences in the end of XIX century, the spectroscopies are essential to the investigation of the structure of the matter. The molecular spectroscopy are grounded in the studies of the transitions between the vibrational and/or rotational levels. Among the techniques that can be used to study the molecular structure, infrared and Raman spectroscopies are in a pivotal position. By using these techniques was possible the determination of structures from dyes [39], metallic complexes [40–42], conducting polymers [43, 44], polymer-clay nanocomposites [15–25] to carbon allotropes [30–38]. In Raman spectroscopy, [45–47] the physical phenomenon is very distinct from the infrared, which is a typical absorption between allowed states, in the case of Raman; there is a scattering process of the incident radiation. The radiation source has much more energy than the vibrational transitions, but through the scattering process, it is possible to screen the vibrational levels (see **Figure 2**).

Another possibility in Raman spectroscopy is the use of different laser lines (E_0), as consequence there is the chance to probe electronic levels in addition to the vibrational ones. When the E_0 is equal or near to an electronic transition there is an increase of the Raman cross-section for at least 10^5 times and also the intensification of the vibrational modes associated to the chromophore structure. The use of microscopies coupled to the Raman instrument permits the investigation of the sample at microscopic level (or nanoscopic level if an electronic or probe microscopies were used) in a non-destructive manner. The main advantage however is the ability to focus the laser on a very small part of the sample (1 μm approximately or smaller). The high lateral resolution and depth of field (the order of a few micrometers) are very useful for the study of multilayered polymeric thin films or others complex materials, such as polymer-clay nanocomposites. In fact, the major part of our studies were conducted in a Raman instrument coupled with an optical microscopy.

1.2.2 X-ray absorption spectroscopy

There are many spectroscopic techniques employed routinely in clay science research in order to investigate multiple aspects of the samples. X-ray spectroscopy has a unique capability to obtain atom-specific information just by tune the correct incident energy of a synchrotron radiation ring. Hence, it is possible to study different atoms and their environments in a clay material or any other complex sample. An X-ray absorption spectrum (XAS) is a consequence of the excitations of a core electron to molecular unoccupied states (or extended states in a case of solid

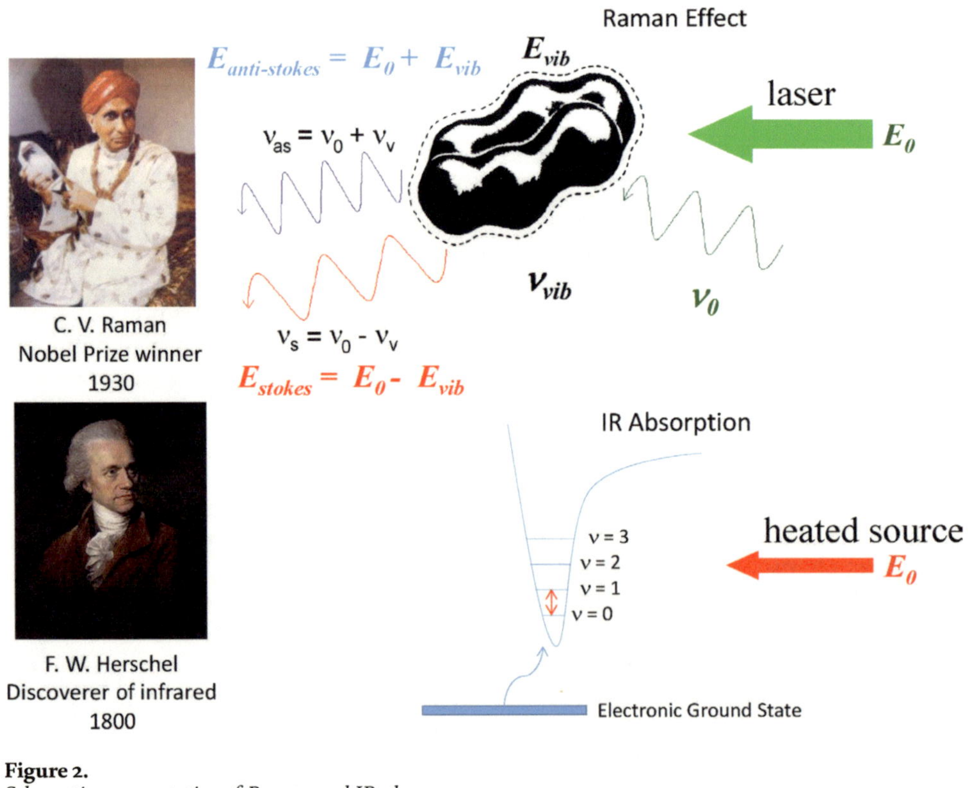

Figure 2.
Schematic representation of Raman and IR phenomenon.

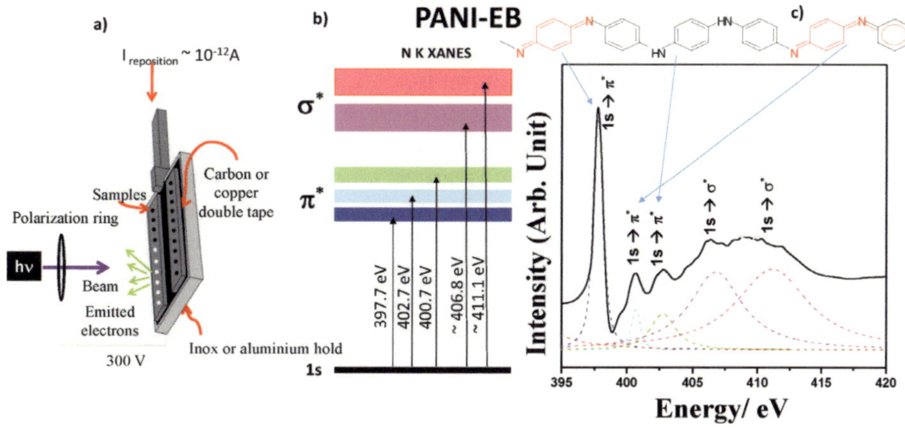

Figure 3.
a) the N K XANES measurements are done in ultrahigh vacuum (the pressure inside the chamber is ca. 10^{-7} mbar). The measured signal is the current of reposition of electrons from the sample (near 10^{-12} a), after the X-ray absorption there are many emission effects (photoelectrons, electrons auger and secondary electrons) those are proportional to the absorption intensity [25]. The arrangement used in our experiments is displayed above. A grooved rod is placed over the main rod to delimit the area (ca. 0.2 cm^2) and prevent the mixing of the samples, since the measurements are made with the rod positioned vertically in the sample chamber. b) Schematic representation of the transitions from 1 s to π^* and σ^* states of a molecular material (PANI-EB), an the correspondent N K-edge X-ray absorption spectrum is also displayed. c) N K XANES spectrum of powered sample of PANI-EB is represented by the black top continuous line (—). The five Voigt bands used in the deconvolution of the experimental spectrum are shown below the experimental data (red dashed lines, – – –). The sum spectrum of the five Voigt bands is also displayed (red pointed line, · · ·). Scheme of PANI-EB is also shown at the top of the figure; the main peaks assigned to 1 s $\rightarrow \pi^*$ transitions are also indicated. Deconvolution of the experimental N K XANES spectrum was done using the (SPSS, 1995) with Voigt bands (Voigt area mode with varying widths) and linear baseline (linear, D2 mode). N K XANES spectrum was obtained in a spherical grating Monochromator (SGM) beam line (dipole magnetic field of 1.65 T and critical energy of 2.08 keV) at the Brazilian National Synchrotron Light Laboratory (LNLS, electron energy of the storage ring of 1.37 GeV). This line can operate in the energy range from 250 eV to 1000 eV, which covers the K edges of carbon (277 eV), nitrogen (392.4 eV), and oxygen (524.9 eV). The SGM beam line has a focused beam of roughly a 0.5 mm^2 spot size with spectral resolution $E/\Delta E$ better than 3.000 and the spectra were recorded in the total electron yield (TEY) with the sample compartment pressure of 10^{-6} Pa. The TEY detection can be briefly described as follows: I(replacement current of electrons) $\propto$ I(emitted electrons) $\propto$ I(absorbed electrons).

samples). For instance, in **Figure 3** is schematically represented the absorption of an N K shell electrons (1 s level) of an atom bonded in a solid material. The absorption occurs if the incident photon energy is transferred to an electron strongly bounded to the atom with sudden changes in the absorption coefficient. The X-ray absorption spectra can be also used for analytical purposes, because the energy edges are characteristic of each chemical element [48–50].

Our group has been used X-ray spectroscopy to investigated different conjugated systems, [16, 17, 23, 25] such as polymers and dyes and their nanocomposites with clays and other materials. The N K-edge XANES spectrum of PANI in its emeraldine base form (EB) is dominated by 1 s $\rightarrow \pi^*$ transitions whose energy values and intensities are related to the oxidation and doping states of PANI (see **Figure 3**). The use of multiple edges permit to probe the polymer or the clay structures such as in the case of polymer-clay nanocomposites.

2. Example of recent investigated system

2.1 Polyaniline-clay materials under heating

Our group have been studied conducting polymer-clay nanocomposites a more than a decade; the main reason is to correlate the electrical and thermal properties of the material with the structural backbone and molecular arrangements of the interlayer polymer. The bulk properties of a conjugated polymer is correlated to the arrangement of its chains [51–53]. By intercalation into clays, it is possible to increase the polymer properties by changing its molecular arranging, but there is also an improvement of properties by interaction with the clay layers. The all reasons for the polymer-clay synergism is not yet completely understood, however many data was acquired in the literature for many polymer layered materials [54–56]. Nanocomposites formed with inorganic host structures and polyaniline and its derivatives have been one of the most studied systems. Among the inorganic hosts employed to confine conducting polymers, clays are frequently used. Our group, have been dedicate much effort to study such system by using mainly resonance Raman and X-ray spectroscopies as the main technique.

Our studies of the structure of PANI intercalated into MMT layers obtained by polymerization in aqueous suspension has modified-JGB-like units (*m*-JGB) in its backbone (see **Figure 4**) [15, 16, 24]. This result was very important in the literature because clearly shown that only using more conventional techniques, such as FTIR and EPR, is not possible to infer conclusively the exact nature of the PANI structure. In addition, our resonance Raman and X-ray absorption studies showed that intercalated PANI has a different chemical backbone than the conventional polymer (free PANI in its emeraldine salt state). More recently, we are interested to investigate the changes of the molecular structure of intercalated anilinium into montmorillonite clay (An^+-MMT) during heating treatment. **Figure 5** shows the resonance Raman spectra of An^+-MMT (see ref. [15, 16] for the description of synthesis of An^+-MMT) under heating at 100°C as a function of time. It is possible to see that the relative intensities of bands related to polaronic units of PANI at ca. 1180 and 1340 cm^{-1}, and also the bands related to cross-linked phenazinic segments at 580, 1380, and 1645 cm^{-1} increase as the time under heating also increases. These changes can be associated to the polymerization of intercalated An^+ without the use of external oxidant, like ammonium persulfate. After one day, there is no more changing.

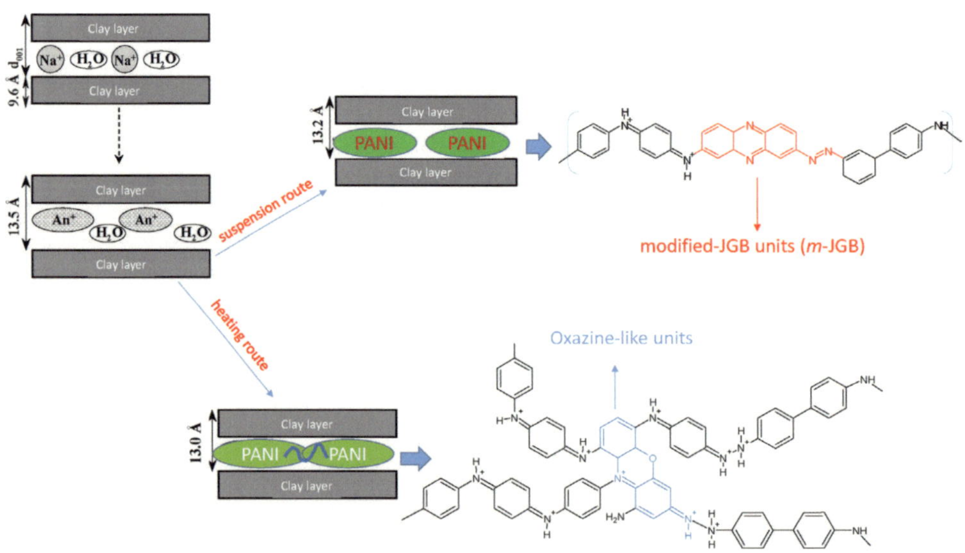

Figure 4.
Schematic representation of intercalation of anilinium (an^+) ion into MMT clay layers and their following polymerization in two different routes. XRD patterns and d_{001} values of powdered samples were obtained on a Rigaku diffractometer model Miniflex using Cu Kα radiation (1.541 Å, 30 kV, 15 mA, step of 0.05°). The possible molecular structures of intercalated PANIs are also displayed.

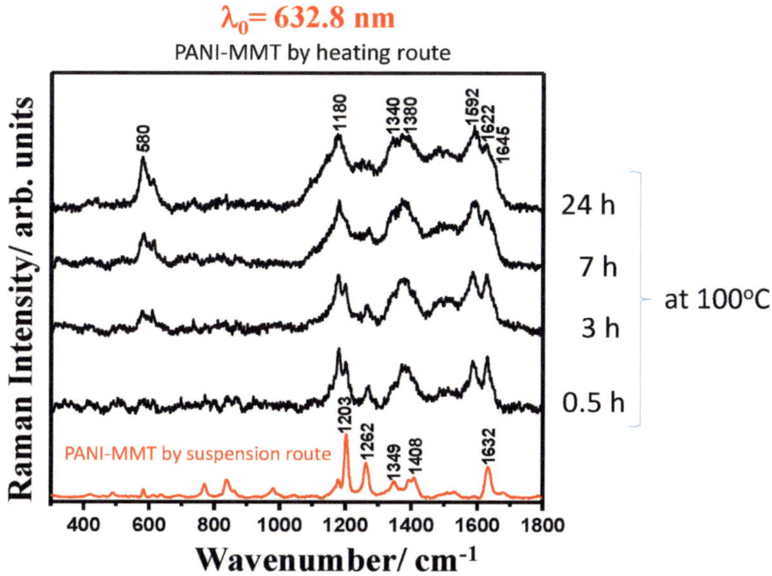

Figure 5.
Resonance Raman spectra of PANI-MMT nanocomposite obtained by suspension route and the an⁺-MMT material under different times (indicated in the figure) of heating at 100°C in air inside an oven. All experiment was done at 100°C. the spectra were acquired in a Renishaw Raman system 3000 equipped with a CCD detector and an Olympus microscope. The laser beam was focused on sample by a 50x lens. Laser power was always kept below 0.7 mW at the sample in order to avoid laser-induced sample degradation. The experiments were performed under ambient conditions using a back-scattering geometry. The samples were irradiated with 632.8 nm ($E_0 = 1.96\,eV$) line of a He–Ne laser.

Hence, it must to emphasize that the RR spectrum of PANI-MMT prepared by heating treatment (spectrum at 24 h) is completely different to the PANI-MMT prepared by *in situ* polymerization in aqueous suspension (spectrum in red). The characteristics bands related to Janus green-like (JGB) units (1203, 1408, and 1632 cm^{-1}) are not observed in the spectrum of PANI-MMT prepared by heating. Hence, the possibility to obtain a PANI only by heating is also very important, however the results clearly show that the polymer obtained in this route is also different from the PANI-MMT obtained from suspension route (see **Figure 4**). This new result clearly demonstrated the great potential of the resonance Raman to the elucidation of structures of intercalated polymers into clay nanocomposites.

We also have recently studied the thermal effects over the structure of PANI-MMT nanocomposites. **Figure 6** shows the resonance Raman spectra of PANI-MMT nanocomposites obtained by suspension route and submitted to heating process in air atmosphere at indicated temperatures. The samples were irradiated with 632.8 nm (E_0 = 1.96 eV) and 488.0 nm (E_0 = 2.54 eV) laser lines. The first thing to be considered is that PANI-MMT nanocomposites showed signal up to 300°C (similar behavior was observed for *in situ* Raman measurements during heating [24]).

The bands related to the *m*-JGB (see **Figure 4** and Raman spectra at 632.8 nm in **Figure 6**) groups are seen up to 200°C, at higher temperatures the bands at 1201, 1412 and 1630 cm^{-1} (related to the *m*-JGB segments) decrease, but at the same time, the bands at 574, 1401, and 1620 cm^{-1} assigned to Oxazine-like units (cross-linking segments, see **Figure 4**) increase. At 488.0 nm, a similar behavior is observed; however only the band at 1401 cm^{-1} associated to Oxazine-like units is clearly seen. This behavior can be rationalized considering that *m*-JGB units are degraded at higher temperatures and the cross-linked PANI-ES bands prevails, it shows that the intercalated polymer has higher thermal resistance than observed in free PANI-ES and PANI-MMT composites.

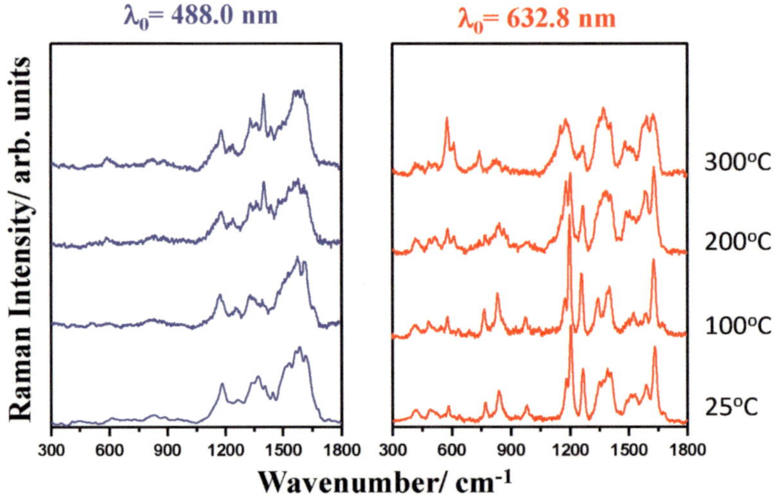

Figure 6.
Resonance Raman spectra of PANI-MMT nanocomposite obtained by suspension route and submitted of heating at indicated temperature in air inside an oven. The spectra were acquired in a Renishaw Raman system 3000 equipped with a CCD detector and an Olympus microscope. The laser beam was focused on sample by a 50x lens. Laser power was always kept below 0.7 mW at the sample in order to avoid laser-induced sample degradation. The experiments were performed under ambient conditions using a back-scattering geometry. The samples were irradiated with 632.8 nm $(E_0 = 1.96\,eV)$ *line of a He–Ne laser and 488.0 nm* $(E_0 = 2.54\,eV)$ *line of an* Ar^+ *laser.*

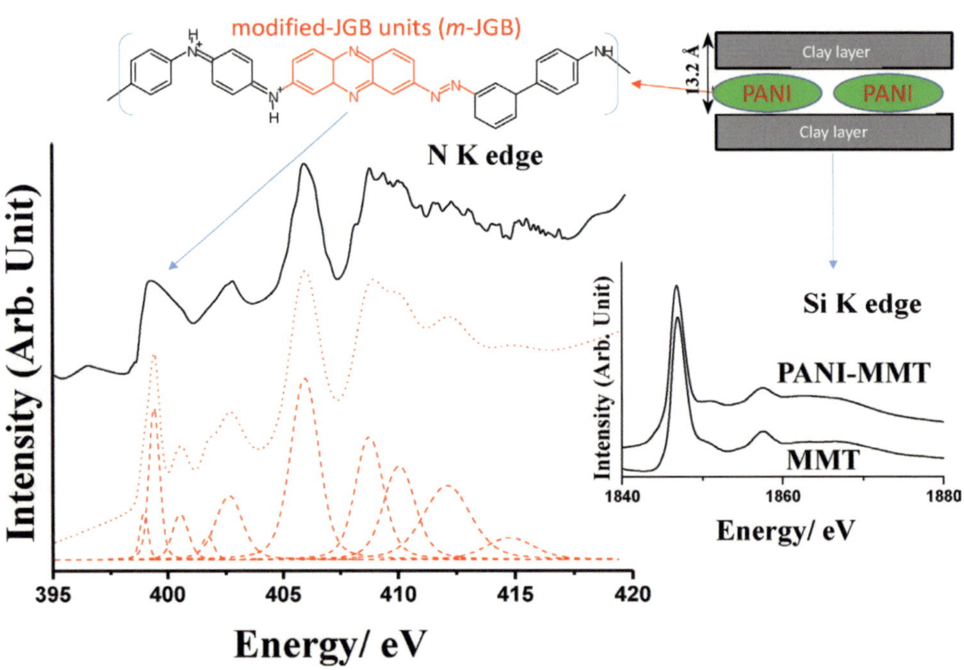

Figure 7.
N K XANES spectra of powered samples of PANI-MMT nanocomposites. The black top continuous line represents experimental spectra. The Voigt bands used in the deconvolution of the experimental spectrum are shown below the experimental data (red dashed lines, – – –). The sum spectrum of the Voigt bands is also displayed (red pointed line, · · ·). The Si K XANES spectra of powered samples of MMT and PANI-MMT are also shown inside the figure. Silicon K-edge spectra were recorded using the total electron yield detection and the samples chamber at ca. 10^{-6} *Pa. All energy values in the Si K-edge spectra were calibrated using the first resonant peak in the Si K XANES spectra for monocrystalline silicon.*

The X-ray absorption studies permit the study of polymer and clay at same time just by selection of appropriate photon energy to probe a specific atomic edge.

Our group have been studied a lot of nitrogen and silicon contend compounds in order to understand to the influence of the chemical structures and its environments over the atomic edges values (mainly Nitrogen, Carbon and Silicon). For instance, **Figure 7** shows the XANES spectra at N and Si K edges of the PANI-MMT nanocomposites. The N K edge gives many peaks related to the complex conjugated structure of the PANI. However, at Si K edge the spectra are simpler due to the regularity of silicon sites into clay layers and by the small influence of intercalated polymer over the electronic properties of clays.

3. Conclusion and future remarks

The screening of the electronic and vibrational structure of polymer-clay nanocomposite through resonance Raman and X-ray absorption spectroscopies has been decisive in determination of their structure and in the study of the interactions between the clays and intercalated polymers in a myriad of synthetic conditions. In fact, by selecting the appropriate photon energies it is possible to study in particular the specific segment of the polymer or clay. The new Raman instruments and new synchrotron rings can give better spectroscopic data associated to a very higher spatial resolution. This open the possibility to study localized inhomogeneity, specific chemical modifications and many other aspects of these extraordinary materials derived from clays.

Author details

Gustavo Morari do Nascimento
Federal University of ABC, CCNH, Santo André, Brazil

*Address all correspondence to: gustavo.morari@ufabc.edu.br; morari@yahoo.com

References

[1] The Holy Bible, English Standard Version. ESV® Text Edition: 2016. Copyright © 2001 by Crossway Bibles, a publishing ministry of Good News Publishers.

[2] Bergaya F, Lagaly G. General introduction: Clays, clay minerals, and clay science. In: Handbook of Clay Science, Bergaya, F. Theng, B. K. G. & Lagaly, G. (Eds.). 1-18 Elsevier, Amsterdam. 2006.

[3] Hall PL. Clays: their significance, properties, origins and uses. In: A Handbook of Determinative Methods in Clay Mineralogy, Wilson, M.J. (Ed.). 1-25, Blackie, Glasgow. 1987.

[4] Guggenheim S, Martin RT. Definition of clay and clay mineral: joint report of the AIPEA nomenclature and CMS nomenclature committees. Clays and Clay Minerals 1995;43: 255 and Clay Minerals 1995;30:257.

[5] Moore DM, Reynolds RC Jr. X-ray Diffraction and the Identification and Analysis of Clay Minerals, 2nd edition. Oxford University Press, Oxford, 1997.

[6] Brown G. Associated minerals. In: Crystal Structures of Clay Minerals and their X-ray Identification. Brindley, G.W. & Brown, G. (Eds.). 361-410, Mineralogy Society, London. 1980.

[7] Yariv S. Introduction to organo-clay complexes and interactions. In: Organo-Clay Complexes and Interactions. Yariv, S. & Cross, H. (Eds.). Marcel Dekker, Inc. New York. 2002.

[8] Wilson, H. Ceramics-Clay Technology. McGraw-Hill Book Co. New York. 1927.

[9] Reis H. Clays, Their Occurrence Properties and Uses. John Wiley & Sons. Committee on Definition of the Term "Ceramics" (1920). Report of the committee. Am. Ceramic Soc. Jour., 1927;3:526.

[10] Norton FH. Elements of Ceramics. Cambridge, MA., Addison-Wesley Press. 1952.

[11] Ozin GA, Nanochemistry – synthesis in diminishing dimensions, Adv. Mater., 1992;4:612.

[12] Do Nascimento GM, Souza MA. Spectroscopy of nanostructured conducting polymers. In: Nanostructured Conducting Polymers. Eftekhari. A.; (Org.). Londres: Wiley and Sons, 341-375. 2010.

[13] E. Ruiz-Hitzky and P. Aranda, Confinement of conductive polymers into inorganic solids, Anales de Quımica Int. Ed., 1997;93:197.

[14] Pavlidou S, Papaspyrides CD. A review on polymer-layered silicate nanocompopsites. Prog. Polym. Sci. 2008;32:1119.

[15] Do Nascimento GM, Constantino VRL, Temperini MLA. Spectroscopic characterization of a new type of conducting polymer–clay nanocomposite. Macromolecules 2002;35:7535.

[16] Do Nascimento GM, Landers R, Constantino VRL, Temperini MLA. Aniline polymerization into montmorillonite clay: a spectroscopic investigation of the intercalated conducting polymer. Macromolecules 2004;37:9373.

[17] Do Nascimento GM, Constantino VRL, Temperini MLA. Spectroscopic characterization of doped poly(benzidine) and its nanocomposite with cationic clay. J. Phys. Chem. B 2004;108:5564.

[18] Do Nascimento GM, Landers R, Constantino VRL, Temperini MLA.

Spectroscopic characterization of polyaniline formed in the presence of montmorillonite clay. Polymer 2006;47:6131.

[19] Do Nascimento GM, Barbosa PSM, Constantino VRL, Temperini MLA. Benzidine oxidation on cationic clay surfaces in aqueous suspension monitored by in situ resonance Raman spectroscopy. Colloids Surf. A: Physicochem. Eng. Aspects 2006;289:39.

[20] Do Nascimento GM, Padilha ACM, Constantino VRL, Temperini MLA. Oxidation of anilinium ions intercalated in montmorillonite clay by electrochemical route. Colloids Surf. A: Physicochem. Eng. Aspects 2008;318:245.

[21] Do Nascimento GM, Temperini MLA. Structure of polyaniline formed in different inorganic porous materials: A spectroscopic study. Eur. Polym. J. 2008;44:3501.

[22] Do Nascimento GM, Temperini MLA. Spectroscopic study of the polymerization of intercalated anilinium ions in different montmorillonite clays. J. Mol. Struct. 2011;1002:63.

[23] Do Nascimento GM. X-ray absorption spectroscopy of nanostructured polyanilines. Chemical Papers. 2013;67:933.

[24] Do Nascimento GM, Pradie NA. Deprotonation, Raman dispersion and thermal behavior of polyaniline–montmorillonite nanocomposites Synthetic Metals 2016;217:109.

[25] Do Nascimento GM. Structure of Clays and Polymer–Clay Composites Studied by X-ray Absorption Spectroscopies. In: Do Nascimento GM, editor. (Org) Clays, Clay Minerals and Ceramic Materials Based on Clay Minerals 1st ed. London: InTech; 2016.

[26] Do Nascimento GM, Silva JEP, De Torresi SIC, Santos PS, Temperini MLA. Spectroscopic Characterization of the Inclusion Compound Formed by Polyaniline and β-Cyclodextrin. Mol. Cryst. Liq. Cryst., 2002;374:53.

[27] Do Nascimento GM. Spectroscopy of Polyaniline Nanofibers. In: Kumar A, editor. (Org.)Nanofibers. 1st ed. Austria/Croacia: InTech; 2010.

[28] Do Nascimento GM. Resonance Raman of Polyaniline Nanofibers In: Michaelson L, editor. (Org.) Advances in Conducting Polymers Research. 1st ed. New York: Nova Publishers; 2014.

[29] Do Nascimento GM. Raman dispersion in polyaniline nanofibers. Vibrational Spectroscopy. 2017;90:89

[30] Dresselhaus MS, Dai H. Carbon Nanotubes: Continued Innovations and Challenges MRS Bulletin 2004;29:237.

[31] Saito R, Fujita M, Dresselhaus G, Dresselhaus MS. Electronic structure of graphene tubules based on C_{60}*Phys. Rev. B* 1992;46:1804.

[32] Dresselhaus MS, Dresselhaus G, Avouris Ph. *Carbon Nanotubes: Synthesis, Structure, Properties and Applications,* Inc: Springer-Verlag, Berlin, 2001.

[33] Dresselhaus MS, Dresselhaus G, Jorio A, Souza Filho AG, Pimenta MA, Saito R. Single Nanotube Raman Spectroscopy *Acc. Chem. Res.* 2002;35:1070.

[34] Do Nascimento GM, Hou T, Kim YA, Muramatsu H, Hayashi T, Endo M, Akuzawa N, Dresselhaus MS. Double-Wall Carbon Nanotubes Doped with Different Br2 Doping Levels: A Resonance Raman Study Nano Lett. 2008;8:4168.

[35] Do Nascimento GM, Hou T, Kim YA, Muramatsu H, Hayashi T, Endo M, Akuzawa N, Dresselhaus MS.

Comparison of the Resonance Raman Behavior of Double-Walled Carbon Nanotubes Doped with Bromine or Iodine Vapors. J Phys. Chem. C 2009;113:3934.

[36] Do Nascimento GM, De Oliveira RC, Pradie NA, Lins PRG, Worfel PR, Martinez GR, Di Mascio P, Dresselhaus MS, Corio P. Single-wall carbon nanotubes modified with organic dyes: Synthesis, characterization and potential cytotoxic effects J. Photochem. Photobiol. A Chem. 2010;211:99.

[37] Do Nascimento GM, Hou T, Kim YA, Muramatsu H, Hayashi T, Endo M, Akuzawa N, Dresselhaus MS. Behavior of the high frequency Raman modes of double-wall carbon nanotubes after doping with bromine or iodine vapors. Carbon 2011;49:3585.

[38] Do Nascimento GM, Barros WP, Kim YA, Muramatsu H, Hayashi T, Endo M, Pradie NA, Fantini C, Pimenta MA, Dresselhaus MS, Stumpf HO. Single-wall carbon nanotube interactions with copper-oxamato building block of molecule based magnets probed by resonance Raman spectroscopy J. Raman Spectrosc. 2012;43:1951.

[39] Bonancea CE, Do Nascimento GM, De Souza ML, Temperini MLA, Corio P. Substrate development for surface-enhanced Raman study of photocatalytic degradation processes: Congo red over silver modified titanium dioxide films Applied Catalysis B: Environmental 2006;69:34.

[40] Silva IF, Teixeira IF, Barros WP, Pinheiro CB, Ardisson JD, Do Nascimento GM, Pradie NA, Teixeira APC, Stumpf HO. An FeIII dinuclear metallacycle complex as a size selective adsorbent for nitrogenous compounds and a potentially effective ammonia storage material. J. Mater. Chem. A 2019;7:15225.

[41] Silva IF, Do Pim WD, Teixeira IF, Barros WP, Teixeira APC, Do Nascimento GM, Pereira CLM, Stumpf HO, Selective Wrapping of Few-Walled Carbon Nanotubes by a Serpent-Like Heterobimetallic Coordination Polymer J. Phys. Chem. C 2016;120:1245.

[42] Mendonça FG, Silva TG, Do Nascimento GM, Stumpf HO, Mambrini RV, Do Pim WD. Human Hair as Adsorbent of Palladium(II) in Solution: A Precursor of Well-Dispersed Size-Controlled Pd Nanoparticles. J. Braz. Chem. Soc., 2019;30:736.

[43] Do Nascimento GM, Silva CHB, Temperini MLA. Spectroscopic characterization of the structural changes of polyaniline nanofibers after heating. Polymer Degradation and Stability. 2008;93:291

[44] Do Nascimento GM, Kobata PYG, Temperini MLA. Structural and vibrational characterization of polyaniline nanofibers prepared from interfacial polymerization. The Journal of Physical Chemistry. B. 2008;112:11551

[45] D. N. Batchelder, In Optical Techniques to Characterize Polymer Systems; H. Brässler, Eds.; Elsevier: Amsterdam, 1987.

[46] D. N. Batchelder, D. Bloor. In Advances in Infrared and Raman Spectroscopy, Wiley-Heyden: London, 1984.

[47] Clark JH, Dines TJ. Resonance Raman Spectroscopy, and Its Application to Inorganic Chemistry. New Analytical Methods. Angew. Chem. Int. Ed. Engl. 1986;25:131.

[48] Margaritondo G, Introduction to Synchrotron Radiation, Oxford University Press, New York, 1988.

[49] Bilderback DH, Elleaume P, Weckert E. Review of third and next

generation synchrotron light sources, J. Phys. B: At., Mol. Opt.Phys. 2005;38:S773.

[50] Bianconi A. In Chemical Analysis; Koningsberger, D. C. & Prins, R., Eds.; John Wiley & Sons: USA, 92, Chapter 11, page 573, 1988.

[51] Shirakawa H. The discovery of polyacetylene film: The dawning of an era of conducting polymers (Nobel Lecture). Angewandte Chemie, International Edition. 2001;40:2575

[52] Liepins R, Ku CC. Electrical Properties of Polymers, C. H. Verlag, Munich, 1987.

[53] MacDiarmid AG, 'Synthetic Metals': a novel role for organic polymers (Nobel Lecture). Angew. Chem. Int. Ed. 2001;40:2581.

[54] Lee D, Char K, Lee SW, Park YW, Structural changes of polyaniline/montmorillonite nanocomposites and their effects on physical properties, J. Mater. Chem., 2003;13:2942-2947.

[55] Wu Q, Xue Z, Qi Z, Hung F, Synthesis and characterization of PAn/clay nanocomposite with extended chain conformation of polyaniline, Polymer, 2000;41:2029-2032.

[56] Zeng QH, Wang DZ, Yu AB, Lu GQ, Synthesis of polymer-montmorillonite nanocomposites by in situ intercalative polymerization Nanotechnology, 2002;13:549-553.

Chapter 2

Clay Hybrid Materials

Tanushree Choudhury

Abstract

The modern trend is to prepare hybrid material using nano clay. Formation of nano clay, an exfoliated clay, and proper dispersion in a polymer matrix remains a challenge. The green composite so formed by clay polymer mixing has many improved properties such as high Tg (glass transition temperature), high flame resistance, high tensile strength, and improved barrier properties, which may find application in textile industry, automobile industry, environmental and polymer engineering field.

Keywords: nano clay, dispersion, CTAB, organic modifier, properties

1. Introduction

Hybrid Materials, in general, have enhanced properties to their components alone. Some of the properties of these hybrid materials, which have been studied in depth, are moduli, thermal expansion coefficients, gas permeability, ionic conductivity, etc. These hybrid materials are classified based on their interaction between host and guest phases. Depending on the type of matrix and guest phase, hybrid materials have been classified into three groups: (i) "OI" organic-inorganic or molecular hybrids, "IO" inorganic-organic intercalation compounds, nanocomposite materials, and solid-state hybrids exhibited by clay-calixarene derivatives [1].

One of such hybrid materials is clay-based hybrid material. Clay minerals are aluminosilicates. Though different types of clay have been used, for making hybrid materials, one of the most commonly used clay is montmorillonite. It belongs to 2:1 type of clay, two silicate layers and one octahedral brucite type of layer containing a mostly aluminum-oxygen hydroxyl group. Isomorphous substitution of trivalent Al^{3+} ion by divalent/monovalent or tetrahedral Si with trivalent Al^{3+} ion leads to charge imbalance in the crystal. This imbalance is compensated by the presence of counter ions present at the surface of the sheet layer. The edge of each platelet has a hydroxide group allowing it to form water gels.

Clay particles are small in size <2 μm, have a large surface area-to-mass ratio. The counter ions (at the exchangeable sites on clay) along with water molecules also serve a bridge between the two layers keeping them inbound.

Exchangeable cations adsorbed on the surface layer can be replaced by other materials. Inherently clay surfaces are hydrophilic attracting polar groups. However, they can be made oleophilic by exchanging the cations with organic ions like cetyl trimethyl ammonium bromide ions or cetyl trimethyl ammonium pyridinium ions etc., producing organoclay composites or polymer clay nanocomposites (PCNs). These composites are extremely investigated in material science and find widespread applications as adsorbents for heavy metal ions [2–4], ceramics and thin films [5], building materials [6], photocatalysts for wastewater treatment [7, 8], drug delivery vehicle [9], bio-inspired materials [10], optoelectronic devices [11], ferrofluids [12], and hydrogel clay hybrids for pesticide and nutrient retention [13].

This review focuses on the properties of the different types of clay hybrid materials that can be prepared by intercalation chemistry, in situ polymerization and sol-gel techniques. It would also provide an insight into the application of these hybrids for a sustainable environment.

2. Clay-based hybrid materials

2.1 How clay can be used in a hybrid material

2.1.1 Structure of clay

Clay minerals belong to phyllosilicates. The principal building elements of the clay minerals are two-dimensional arrays of Si-O- tetrahedral and 2-D arrays of Al or Mg-O-OH octahedral as shown in **Figure 1**. In most clay minerals, such sheets of tetrahedral and octahedral are superimposed in different fashions [14].

a. **Structure of tetrahedral sheet:** In the Si-O sheets, the Si atoms are coordinated with four oxygen atoms. The O atoms are located on the four corners of a regular tetrahedron with the Si atom in the center. In the sheet, three of the four oxygen atoms of each tetrahedron are shared by three neighboring tetrahedral. The fourth oxygen atom of each tetrahedron is pointed downward. This Si-O sheet is called a tetrahedral sheet or silica sheet.

b. **Structure of octahedral sheet:** In the Al, Mg-O-OH sheets, the Al or Mg atoms are coordinated with six oxygen atoms or –OH groups, which are located around the Al or Mg atom, with their centers on the corners of a regular octahedron resulting in hexagonal close packing. This sheet is called alumina or magnesia sheet. The fourth oxygen atom protruding from the tetrahedral sheet is shared by the octahedral sheet. This sharing of atoms may occur between one silica and one alumina sheet as 1:1 layer minerals.

In 2:1 layer minerals, one alumina or magnesia sheet shares oxygen atoms with two silica sheets, one on each side. These layers in clay minerals are stacked parallel to each other.

2.2 Origin of surface charge and modification of clay surface

In the tetrahedral sheet, tetravalent Si is sometimes partly replaced by trivalent Al. In the octahedral sheet, there may be replacement of trivalent Al by divalent Mg without complete filling of the third vacant octahedral position. Al atoms may also be replaced by Fe, Cr, Zn, Li, and other atoms. The small size of these atoms permits them to take the place of small Si and Al atoms; therefore, the replacement is often referred to as isomorphous substitution. When an atom of lower positive charge replaces one of higher valence, a deficit of positive charge takes place or in other words, excess of negative charge. This excess of negative charge is compensated by the adsorption on the layer surfaces of cation, which are too large to be accommodated in the interior of the crystal. In the presence of water, the compensating cations on the layer surfaces may be easily exchanged by other cations when available in solution, hence they are called exchangeable cations. Thus clay minerals bear the potential of forming hybrid materials.

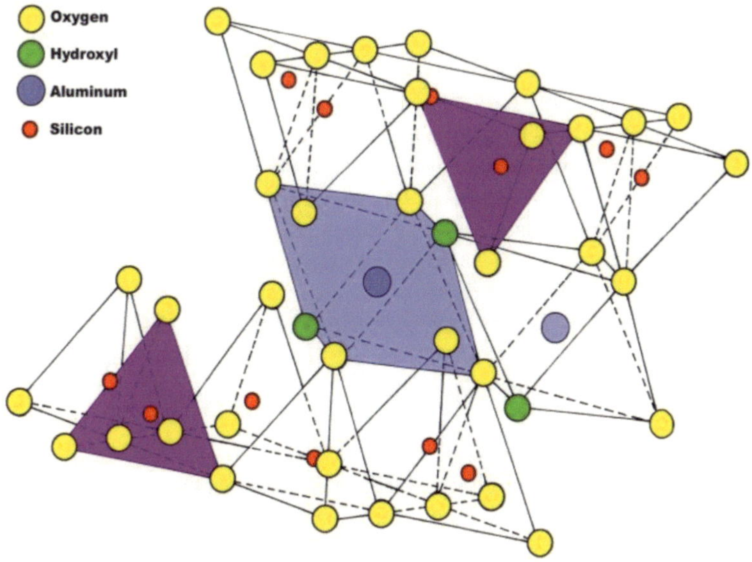

Figure 1.
Structure of clay mineral.

Natural clay is hydrophilic. The surface of the clay needs to be modified so that it can interact with hydrophobic polymers. The modification of clay surface is generally done by the cation exchange process. The ability of clays to exchange cations between each of their layers and retain them is a unique characteristic. The intercalated cations can be exchanged by other cations by treatment of other cations in solution. This cation exchange capacity can be defined as the maximum amount of cations that a given amount of clay can take up and this is constant. The ability to cation exchange in the interlayer space determines the adsorption ability of montmorillonite [15]. The most exchangeable cations that can be adsorbed on the clay surface by the cation-exchange process are inorganic ions (mostly Na^+, Ca^{2+}, Mg^{2+}, K^+), amines, amino acids, cationic surfactants, and non-ionic surfactants. The surface of the clay can also be rendered organophilic by the reaction of hybrid monolayers of clay mineral and amphiphilic alkyl amino cation using Langmuir-Blodget method [16]. When a solution of an amphiphilic alkyl ammonium cation is spread onto clay suspension, negatively charged clay platelets in the suspension are adsorbed onto a floating monolayer of the alkylammonium cation at an air-clay suspension interface. The hybrid monolayers of clay platelets and alkylammonium cations formed at the interface can be transferred onto a solid surface to fabricate a hybrid multilayer.

2.3 Types of clay hybrid materials

The common types of clay hybrid materials are:

1. Intercalation compounds

2. Exfoliated/delaminated compounds

3. Sol-gel hybrid materials

Intercalation compounds: These compounds result from the intracrystalline insertion of organic compounds inside the layers of certain lamellar solids as shown in **Figure 2**.

i. Intercalation of ionic species: Clay minerals exhibit isomorphous substitution as a result of which Si is replaced by Al in a tetrahedral layer or Al by Mg in octahedral layer leading to charge deficiency, which in turn is compensated by exchangeable cations. Exchangeable metal ions located in the interlamellar space of MMT may be replaced by different organic cations such as alkylammonium ions in solution. Alkylammonium cations thus incorporated in organosilicates lower the surface energy of inorganic hosts and improve the wetting characteristic with polymer [17]. It provides a functional group that can react with polymer or initiate polymerization of monomers to improve interfacial strength between inorganic host and polymer. Layered double hydroxides (LDH), for example, $Mg_6Al_2(OH)_{16}CO_3$. $4H_2O$, have a positive charge on the Mg $(OH)_2$ layers. They provide an opportunity for intercalation with organic anions [18].

ii. Intercalation of neutral species: Formation of organic-inorganic compounds by intercalation of neutral molecules in 2D solids, generally phyllosilicates of clay minerals family and it is also observed for other inorganic layered materials, for example, 2D transition metal halides, oxyhalides, dichalcogenides, graphite, and graphite oxide, layered phosphates, phosphorous trichalcogenides. Different mechanisms are proposed that come into play during host-guest interactions in these intercalated materials. Van der Waals forces mainly occur when long-chain alkylammonium ions are inserted in the clay layers. When interlayer cations preserve the hydration shell, an association of molecules takes place with water molecule acting as H bond bridges, which on heating, eliminates water and produces direct coordination between guest species and involved transition metal [19]. Macrocyclic compounds such as crown ether and cryptands penetrate the interlayer space of phyllosilicates and other layered solids giving stable intracrystalline complexes [20]. The interlayer environment of certain layered solids exhibits acid character, which is typical of clay minerals group. Interaction with basic species produces a proton transfer between the inorganic host and organic guest molecule. The organic molecules being protonated gives rise to organic cations balancing the electrical charge of the silicate [21]. Many redox reactions occur during intercalation of organic and organometallic species into various 2D solids. Clays containing interlayer cations like Cu (II) can interact with aromatic compounds such as benzene giving intercalation compounds characterized by the existence of σ or π bonds between the host solid and guest molecule [22].

Exfoliated or delaminated compounds: These compounds are formed when the layers of clay are delaminated and the resulting platelets are homogeneously dispersed throughout the polymer matrix as shown in **Figure 3**. The resulting materials

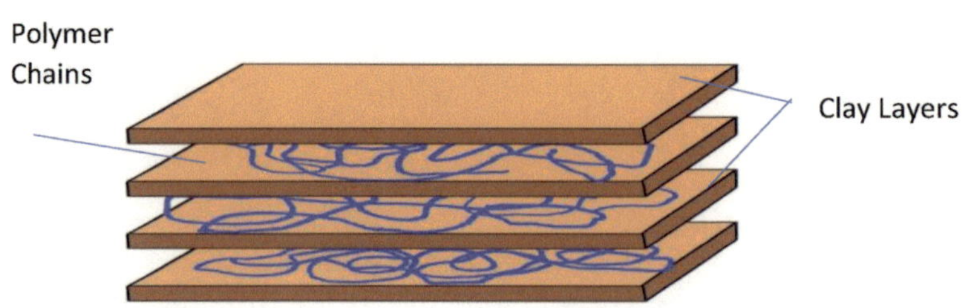

Figure 2.
Intercalated clay.

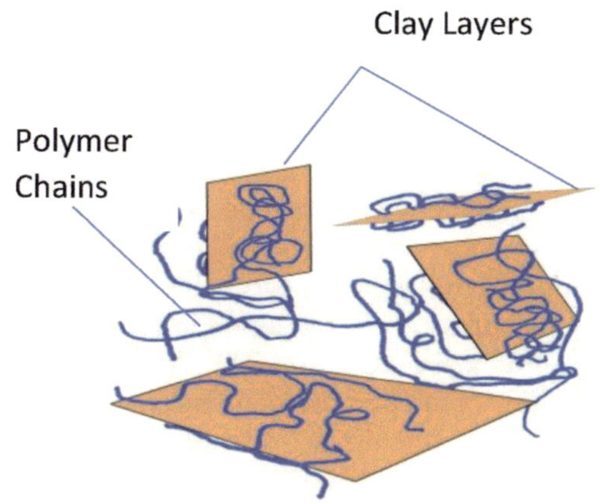

Figure 3.
Exfoliated clay.

are considered as nanocomposites as the interaction takes place at the atomic level between inorganic hosts and organic guest molecules.

Clay mineral is a potential candidate for the filler of these hybrid materials since it is composed of layered silicates, 1 nm thick, which can undergo intercalation with organic molecules [23]. The mechanism of interaction of clay with different polymers is discussed below.

i. Vinyl polymers: These include vinyl addition polymers derived from monomers like methyl methacrylate [24–30], acrylic acid [31], vinyl ester [32], vinyl polymer [33, 34], styrene [35–38], allyl ester resin [39], and acrylates [40].

Studying the mechanism of interaction of ethylene-vinyl acetate with clay, it has been found that as VA (vinyl acetate) content increases, copolymer presents increasing polarity but lower crystallinity with different mechanical behavior. Increasing polarity with increasing VA content is useful in imparting a high degree of polymer-clay surface interaction. Structure and mobility properties of EVA polymer are influenced by VA content and this chain mobility in and around clay galleries tend to modify the level of interaction in clay hybrid materials [41]. Polymer chains of PVA get adsorbed on individual inorganic lamellae in stages after the exfoliation of the clay mineral leading to the formation of intercalated nanocomposites [42]. PVA forms a composite structure with sodium montmorillonite and studies reveal the existence of both exfoliated and intercalated MMT layers for low and moderate silicate loadings. Exfoliation of layers has been attributed to water casting method used since the water suspended layers become kinetically trapped by the polymer and cannot reaggregate [43]. Syndiotactic PS (thermoplastic polymer) differs from other PS (such as a-PS) in that phenyl rings regularly alternate from side to side concerning polymer chain backbone. Two important factors responsible for homogenous dispersion of clay layers in s-PS hybrids are: (a) surfactant should be intercalated between silicate layers of clay by ionic bonding and (b) the hydrophobic tail of the surfactant molecule should be partially compatible or interacted with s-PS molecules [44].

Organophilic modification of clay and amine-terminated PS employing anionic polymerization yielded completely exfoliated hybrids with aspect ratio exceeding 600 when such organoclays were melt compounded with PS. In contrast to this, small molecular weight modifiers only promoted intercalation and failed to exfoliate silicate particles during melt compounding. ABS (thermoplastic polymer) forms an intercalated-exfoliated composite with clay. HENA (hydroxyethyl isonicotinamide) is used as an anchor monomer for homogeneous dispersion of clay minerals in PET (polyethene terephthalate) matrix [45].

Acrylonitrile co-monomer incorporated into poly (styrene-co-acrylonitrile) copolymer accelerates intercalation of copolymers into the galleries of silicate layers modified with an organic intercalant. The faster intercalation of a matrix polymer leads to the better dispersion of silicate layers in the matrix polymer [46]. In the hybrids of SAN, clay particles or nanoscale building blocks are distributed uniformly and their sizes are strongly dependent on co-monomer content. Acrylonitrile co-monomer incorporated into SAN facilitates the intercalation of copolymers into the galleries of silicate layers modified with an intercalant. H-bonding interaction between the nitrile groups of SAN and –OH groups on silicate layer makes a negative contribution to exchange energy of mixing so that the intercalation of copolymers into the galleries of the silicate layer is accelerated. It is also expected that the enhanced polarity of SAN due to incorporated acrylonitrile co-monomer can destroy H-bondings of intercalant in the galleries. This might also increase the rate of intercalation of SAN into the galleries of silicate layers modified with intercalants. Synthesis of hybrid materials using natural clay and modified PVC resulted in the removal of heavy metals (Fe, Cu, Pb, Zn, Cd, Co, and Mn) from aqueous solution and also exhibited good adsorption capacity for Fe (III) [47].

ii. Condensation step polymers: Several important polycondensates have been used in nanocomposites preparation with layered silicates. They are polyamides [48], polyimides [49], polyurethane urea [50], polyurethane [51], poly (butylene terephthalate) [52], poly (ethylene terephthalate co-ethylene naphthalate) [53], epoxy [54, 55], amino, sulfonic acid, and silyl functionalized groups [56–59], and surface-modified groups [60].

Polyamides and polyimides are polymers containing polar functional groups and form homogenous and exfoliated dispersion of silicate layers as silicate layers of clay have polar functional groups and are compatible only with polymers containing polar functional groups. Polyamide-6 (PA6), Polyamide 66 (PA-66), and Nylon form majority of commercial polyamides. PA-66 contains a mixture of chains (only amines, acid groups, or a mixture of two). Differences in end group configuration can lead to significant differences in morphology and properties of blends with functionalized polymers. A lower degree of exfoliation in PA-66 nanocomposites, the affinity of PA-66 for organoclay is less than PA-6 nanocomposites forming the basic aspect of difference in the chemical structure of two polyamides [61]. Nylon-6-clay hybrids (NCHs) have been prepared by using $10A^0$ silicate layers of clay minerals, which are dispersed homogeneously in the polymer matrices resulting in a drastic change in properties (high strength, high modulus, high heat and distortion temperature) and this has been achieved with only a few % of clay [62]. The compatibility of forming hybrids with clay and polymers containing amide and imide groups increases as both contain polar functional groups. PBI (polybenzimidazole) is a thermally stable thermoplastic polymer, contains 1,3-dinitrogen heterocycle. When PBI is added to clay suspensions in a polar environment, the mineral layers will first adsorb the bulky macromolecule and the intercalation proceeds to completion via exchanged sites of the organically modified clay [63]. Polyurethane (PU) elastomers are segmented polymers with soft segments derived from polyols and hard segments from isocyanates and chain extenders. Linear PU is obtained by poly condensate technique using a mixture of diols and diisocyanate. MMT nanolayers are dispersed in PU matrix replacing hydrophilic organic exchange cations of native mineral with more organophilic diethanolamine/triethanolamine. Presence of these groups in galleries of MMT renders them organophilic and promotes the absorption of diol into the interlayer of MMT and improves the particle-matrix interactions since di and triethanolamine contain functional groups, which react with diisocyanate [64].

Pure PU exhibits an amorphous halo at 20° in 2θ. The gallery spacing of the layered clay is 1.1 nm. The gallery spacing of the layered clay in the composites increases to 1.6 nm for the PU/layered nanocomposites. This indicates that PU chains were intercalated between the layers of clay [65]. A multilayered structure consisting of alternate PU chains stacked with the layers of the silicate layers in the microstructure of PU/OMT nanocomposites has been confirmed in literature [66].

With increasing and urgent market demand to produce higher performance electronic devices with a smaller size, lighter weight, and better quality, developing PI films with low coefficient of thermal expansion (CTE) has increasingly become one of the most important issues. The best way to lower the CTE of PI is to introduce low CTE inorganic materials such as clay into PI matrix, yielding PI/clay hybrid composites [67]. Polar aprotic solvents are used for the synthesis of these hybrids, but due to solvent-solute interaction, are not easily removed from the PAA film at temperatures used during thermal curing for PI. The residual solvent causes the PAA (polyamic acid)/clay films to be plasticized during thermal imidization and thus leads to PI/Clay hybrid films with relatively higher values in CTE, but lower than pure film. To eliminate such negative effects of the aprotic solvents on CTE, PAA solutions not containing them should be prepared. A novel PI/clay hybrid film prepared from PAA salt of triethylamine and organoclay in a mixed solvent of THF/MeOH is described in the literature. It is expected that the hybrid will have a much lower CTE than those obtained from PAA in an aprotic solvent.

Full separation of clay layers in the polymer matrix is also achieved by using epoxy resin, which has high polarity and curing property [68]. The presence of polar –OH groups in clay layers impede nonpolar species from entering the galleries and exfoliating the clay layers. The mechanism of clay exfoliation in epoxy clay systems have been studied and reported in the literature. According to it, the acidity of alkylammonium ions catalyze homopolymerization of diglycidyl ether of bisphenol A (DGEBA) molecules inside the clay galleries. CEC of clays determines the number of alkylammonium ions present between clay layers and therefore controls the space available for the diffusion of DGEBA molecules during mixing of organoclay with epoxy resin.

iii. Polyolefins: These include polypropylene [69–74], polyethylene [75–77], ethylene propylene diene methylene linkage rubber [78].

These polymers do not contain any polar groups and homogenous dispersion in the silicate matrix is difficult. Homogenous dispersion of silicate layers in PP is not realized even by using an MMT intercalated with di-stearyl ammonium ion (DSDM-MT) in which polar surfaces of clay are covered with nonpolar long alkyl groups. A novel method of preparing PP-clay hybrid has been developed. PP is mixed with DSDM-MT and polyolefin oligomer with polar telechelic –OH groups (PO-OH) as a compatibilizer. In this process, PO-OH oligomer intercalates between the layers of clay through the strong H-bonding between –OH groups of PO-OH and oxygen groups of silicates. Interlayer spacing increases thus resulting in weakening of interaction between layers.

Another method of preparing PP/clay nanocomposites is by improving the compatibility of PP with organoclay by functionalizing the backbones of PP with polar monomers such as epoxy and maleic anhydride (MA) [79]. Compatibilizers promote compatibility of clay and polymer for good nano dispersion. Polyolefin-graft

MA as compatibilizer is used to enhance the possibility of intercalation of polymer between clay layers. The presence of MA increases the possibility of nanocomposite formation for PS, but this does not appear to help PP. PP/clay nanocomposites modified with the optimum level of compatibilizer yielded the greatest improvement of composite properties [80].

PE (polyethene) is another widely used polyolefin polymers. Alkylammonium ion facilitates interaction with polymer because it renders hydrophilic clay surface organophilic. Organically modified clay is not well dispersed in nonpolar PE as the nonpolar groups are too hydrophobic. Exfoliation and interaction behaviors depended on the hydrophilicity of PE grafted with MA and chain length of the organic modifier in the clay. When the number of methylene groups in alkyl amine (organic modifier) was larger than 16, exfoliated nanocomposites were obtained and the MA grafting levels was higher than about 0.1 wt% for the exfoliated nanocomposite with modified clay [81].

Rubber is another such polymer. Carbon blacks are excellent reinforcers due to their strong interaction with rubbers, but they often decrease the processability of rubber compounds because of high viscosity at high volume loading. MMT exchanged with a liquid rubber (LR) is termed as LR-MMT for utilization of its favorable shape. Co-vulcanization of nitrile rubber was done with LR-MMT for the formation of the molecular composite. It has been studied that there exist strong rubber-filler interactions as (comparable to those in carbon black filled system) in LR MMT, in which negatively charged silicate layers are bonded to LR molecules with positively charged terminal sites forming “bound rubber.”

Polymers grafted on silicate surfaces also helps in delamination of its layers. It has been reported in the literature that PDMS (polydimethylsiloxane) grafted onto MMT layer surface via condensation of hydroxyl groups of PDMS and those hydroxyl groups on MMT layers prevents the nanolayers of MMT from reaggregating.

Intercalation of EPDM chains into OMMT galleries provided a strong interaction between EPDM and OMMT sheets in exfoliated composites.

It has been observed that the photoluminescence quantum efficiency of conjugated polymer PE improves manifold in the presence of the inorganic phase like montmorillonite clay [82]. Incorporation of montmorillonite clays into conjugate polymers like PAni gives rise to hybrid/inorganic composites with special properties for application in organic light-emitting diodes (OLEDs), organic field-effect (OFETs), organic solar cells (OSCs), and electrochromic devices (ECDs) [83].

iv. Fiber-reinforced polymers: Many fibers have been added as reinforcements to the polymer matrix. They are PS-Sisal fiber composites [84]. Bamboo polymer composites [85], short oxide fiber reinforced in kaolin [86], bamboo glass-reinforced in PP [87], thermoplastic starch [88], switchgrass.

v. Biodegradable-polymers: These include biodegradable resin clay composites [89].

Metal incorporated clay composites such as phosphorous clay composites show improved fire performance [90].

Starch modified by grafting with vinyl monomers (e.g., methyl acrylate) onto the starch backbone yielded thermoplastic materials. Kaolin, a natural mineral, hydrated aluminosilicate, with high surface and presence of polar groups showed very good compatibility with thermoplasticized starch.

Aliphatic polyesters, polylactide (PLA) comes under the area of environmentally degradable polymer materials. These are well suited for the preparation of disposable devices because of their biodegradability. The main characteristics of the PLA matrix are its easiness to degrade by the enzymatic or hydrolytic way. Hydrolytic degradation of PLA is a well-known process. Hydrolytic chains cleavage proceeds preferentially in amorphous regions, leading therefore to an increase of polymer global crystallinity. The formation of lactic acid oligomers, which directly follows from this chain scission, increases the –COOH end groups concentration in the medium. These carboxylic functions are known to catalyze the degradation reaction. Relative hydrophilicity of clay plays determining roles in the hydrolytic degradation process. More hydrophilic the filler, more pronounced is the degradation. Thermoplastic corn starch (TPS) clay hybrids showed enhanced biodegradation as compared to TPS alone [91].

vi. Hyperbranched polymers: These polymers have a tree-like structure with a large number of branch points radiating from a multifunctional core molecule and hence a potentially high degree of end-group functionality per molecule. The –OH end groups are assumed to be concentrated in the periphery of the molecules in a hydrophilic environment. Polyester HBPs also show excellent processing char and shrinkage control. Dispersion of HBPs with various types of organically modified MMT in THF led to intercalation over the whole range of MMT contents and the layer expansion correlated with the polarity of organic modifier rather than the size of HBPs [91].

HBPs with highly branched, 3D structure and high concentration of end groups have the promise of good internal bonding rubber phase due to the presence of surface functional groups, in addition to low initial viscosity [92].

Sol-gel hybrid materials: This class of hybrids has received different names such as ORMOSILS and ORMOCERS, referring to organically modified silicates or ceramics, respectively.

The technological importance of the sol-gel process is due to the simplicity in its preparation. Silicon alkoxides are the main precursors used in the synthesis of glasses and ceramics and they are also being used in the preparation of new organic-inorganic hybrid materials [93].

A solution of the molecular precursor is transformed into a sol or a gel by a chemical reaction, resulting in a solid material upon evaporation. This transformation allows the production of materials with different possible compositions, intercalated microstructures, and chemical homogeneity at temperatures less than those used for fusion.

Typical sol-gel processing variables leading to different morphologies of the materials are water to alkoxy and catalyst to alkoxy ratios and the type of solvent and catalyst used [94].

3. Conclusion and outlook

Clay mineral poses a host of technical issues, such as dispersion of the inorganic filler in the polymer/base matrix. Better is the dispersion, better is the hybrid. Clay dispersed in natural dispersant renders the most thermally stable organoclay.

Functionalization of clay surface for better compatibility with polymers is needed for the development of new synthetic layered materials with a wide range of properties. PAni/clay hybrids have been widely studied due to many advantages such as high optical contrast (%T), environmental stability, as well as comparatively low cost. However, the difficulties in processing PAni into films due to its very low solubility in most of the available solvents and the relatively poor mechanical properties decrease its performances and abilities in such applications. Green hybrids reinforced with natural fibers and macromolecules have pronounced biodegradable and recyclable properties and thus emerge as better packaging materials.

Current research focuses on the use of advanced nanotech catalysts and materials for the purification/remediation of contaminated surface or groundwater and municipal water or industrial wastewater. Though clay hybrids have been extensively used as nano adsorbents for the removal of heavy metals, As, and dyes from wastewater, its fabrication as inorganic membranes have received limited attention in the literature. Titania pillared clay, an important class of layered materials, exhibits unique surface charge characteristics that make them a good candidate for removal of organics from wastewater by just adjusting the pH of the solution. Development of such membrane reactors integrating the separation process with photocatalysis would lead to an important new technological application that would add economic value to the vast natural deposits of clay minerals located worldwide. However, membrane fouling is still a critical problem that results in flux decline with time, needs to be addressed. In a nutshell, the outlook is bright and sustainable for clay hybrid materials.

Abbreviations

MMT	montmorillonite
VA	vinyl acetate
EVA	ethyl vinyl acetate
PVA	polyvinyl alcohol
ABS	acrylonitrile butadiene styrene
PET	poly (ethyl terephthalate)
HENA	hydroxyethyl iso-nicotinamide
SAN	styrene-co-acrylonitrile
PA	polyamide
PU	polyurethane
OMT	organic montmorillonite
PI	polyimide
CTE	co-efficient of thermal expansion
PAA	polyamic acid
DGEBA	diglycidyl ether bisphenol A
CEC	cation exchange capacity
DSDM-MT	distearylammonium montmorillonite
PO	polyolefin oligomer
PP	polypropylene
MA	maleic anhydride
PE	polyethylene
LR	liquid rubber
PDMS	polydimethylsiloxane
EPDM	ethylene propylene diene methylene linkage rubber
PLA	polylactide
HBP	hyperbranched polymer
mPE	metallocene blended polyethylene
AN	acrylonitrile
PMMA	polymethyl methacrylate
PPCN	polypropylene clay nanocomposites
OMS	organomethylsilicate
PAM	polyacrylamide
EVOH	ethylvinylalcohol

MAO	methylaluminoxane
TMA	trimethylalkoxide
PAni	polyaniline
PCL	polycaprolactum
PHA	polyhydroxyalkanoate
PBS	polybutylene succinate
DEP	resin
PCN	polymer clay nanocomposites
PS	polystyrene
NCH	nylon clay hybrid
PBI	polybenzoimidazole
OMT	organo montmorillonite
THF	tetrahydrofuran
MeOH	methanol
OLED	organo light emitting diode
OFET	organo field effect
OSC	organic solar cell
ECD	electrochromic device
TPS	thermoplastic corn starch
ORMOSIL	organo modified silicate
ORMOCERS	organo modified ceramics

Author details

Tanushree Choudhury
Chemistry Division, VIT Chennai, Chennai, India

*Address all correspondence to: tanushree.c@vit.ac.in

References

[1] Othman AB et al. Hybrid materials based on clay and calixarenic derivatives. Journal of Porous Materials. 2019;**26**:493-504

[2] Varadwaj GBB, Parida K, Nyamori VO. Transforming inorganic layered monmorillonite into inorganic-organic hybrid materials for various applications. Inorganic Chemistry Frontiers. 2016;**3**:1100-1111

[3] Ray PZ, Shipley HJ. Inorganic nano-adsorbents for the removal of heavy metals and arsenic: A review. RSC Advances. 2015;**5**:29885-29907

[4] Bhatnagar A, Sillanpaa M. Utilization of agro-industrial and municipal waste materials as potential adsorbents for water treatment: A review. Chemical Engineering Journal. 2010;**157**:277-296

[5] Choudhury T. Natural dye intercalated clay hybrid material: Clay based nanopigment. Clay Research. 2017;**36**(1):17-22

[6] Takeda H et al. Characterization of zeolite in zeolite-geopolymer hybrid bulk materials derived from kaolinitic clays. Materials. 2013;**6**:1767-1778

[7] Cristiani C et al. Polyamine-based organo-clays for polluted water treatment: Effect of polyamine structure and content. Polymers. 2019;**11**:897-913

[8] Akkari M et al. Organo clay hybrid materials as precursors of porous ZnO/silica-clay heterostructures for photocatalytic applications. Beilstein Journal of Nanotechnology. 2016;**7**:1971-1982

[9] Kim MH et al. Review of clay-drug hybrid materials for biomedical applications: Administration routes. Clays and Clay Minerals. 2016;**64**(2):115-130

[10] Lin AYM, Meyers MA. Interfacial shear strength in abalone nacre. Journal of the Mechanical Behavior of Biomedical Materials. 2009;**2**:607-612

[11] Silva RC et al. Electrochromic properties of polyaniline based hybrid organic/inorganic materials. Journal of the Brazilian Chemical Society. 2016;**27**(10):1047-1057

[12] Son YH et al. Structure-property correlation in iron oxide nanoparticle-clay hybrid materials. Chemistry of Materials. 2010;**22**(7):2226-2232

[13] Yacoub E. Hydrogel-clay hybrid materials for pesticide and nutrient retention. Laboratoire Mapiem (EA4323) Materiaux Polymeres Interface. Available from: http://mapiem.univ-tln.fr/-PhD-Thesis-.html

[14] Van Olphen H. Clay Colloid Chemistry. New York: Interscience Publishers; 1963. pp. 59-83

[15] Kozak M, Domka L. Adsorption of quarternary ammonium salts on montmorillonite. Journal of Physics and Chemistry of Solids. 2004;**65**(2-3):441-445

[16] Umemura Y. Layered structure of hybrid films of an alkylammonium cation and a clay mineral as prepared by LB method. Thin Solid Films. 2003;**426**:216-220

[17] Sinha Ray S, Okamato M. Polymer/layered silicate nanocomposites. Progress in Polymer Science. 2003;**28**:1539-1641

[18] Riechle WT. Catalytic reactions by thermally activated synthetic anionic clay mineral. Catalysis. 1985;**94**(2):547-557

[19] Yariv S. In: Yariv S, Cross H, editors. Organo-Clay Complexes and Interaction. New York: Marcel Dekker Inc.; 2002

[20] Ruitz-Hitzky E, Casal B. Crown ether intercalations with phyllosilicates. Nature. 1978;**276**:596-597

[21] Fripiat JJ, Chaussiden J, Jelly A. Chimie Physiquedes Phenomenes de Surfaces et Application aux Oxydes et aux Silicates. Paris: Masson; 1970

[22] Doner HE, Mortland MM. Benzene complexes with Cu (II) montmorillonite. Science. 1969;**166**:1406-1407

[23] Okada A, Usuki A. The chemistry of polymer clay hybrids. Materials Science and Engineering. 1995;**C**(3):109-115

[24] Zhang W, Liang Y, Luo W, Fang Y. Effect of clay modifying agents on morphology and properties of PMMA/clay nanocomposites synthesized via gamma ray irradiation polymerization. Journal of Polymer Science Part A: Polymer Chemistry. 2003;**41**:3218-3226

[25] Jash P, Wilkie CA. Effect of surfactants on the thermal and fire properties of PMMA/clay nanocomposites. Polymer Degradation and Stability. 2005;**88**:401-406

[26] Luo JJ, Daniel IM. Characterization and modelling of mechanical behavior of polymer/clay nanocomposites. Composites Science and Technology. 2003;**63**:1607-1616

[27] Stadtmueller LM, Ratinac KR, Ringer SP. The effects of intragallery polymerization on the structure of PMMA-clay nanocomposites. Polymer. 2005;**46**:9574-9584

[28] Zheng X, Jiang DD, Wilkie CA. Methyl methacrylate oligomerically modofied clay and its PMMA nanocomposites. Thermochimica Acta. 2005;**435**:202-208

[29] Zhu J, Start P, Mauritz KA, Wilkie CA. Thermal stability and flame retardancy of PMMA-clay nanocomposites. Polymer Degradation and Stability. 2002;**77**:253-258

[30] Su S, Jiang DD, Wilkie CA. PMMA, PP, PE nanocomposites formation by melt blending. Polymer Degradation and Stability. 2004;**83**:321-331

[31] Onal M, Srikaya Y, Alemdaroglu T, Bozdogan I. Isolation and characterization of a smectite as a micro-mesoporous material from bentonite. Turkish Journal of Chemistry. 2003;**27**:683-693

[32] Kocsis JK, Gryshchuk O, Frohlich J, Mulhaupt R. Interpenetrating vinylester/epoxyresins modified into organophilic layered silicates. Composites Science and Technology. 2003;**63**:2045-2054

[33] Artzi N, Nir Y, Narkis M. Structure and properties of EVOH/clay nanocomposites. Polymer Composites. 2003;**24**(5):627-639

[34] Silva TF et al. Silylated montmorillonite as nanofillers for plasticized PVC nanocomposites: Effect of the plasticizer. Applied Clay Science. 2014;**99**:93-99

[35] Sohn JI, Lee CH, Lim ST, Kim TH, Choi HJ, Jhon MS. Viscoelastic and relaxation characteristic of PS/clay nanocomposites. Journal of Materials Science. 2003;**38**:1849-1852

[36] Yoon JT, Jo WH, Lee MS, Ko MB. Effect of comonomers and shear on the melt intercalation of styrenics/clay nanocomposites. Polymer. 2001;**42**:329-336

[37] Zhang G, Li YJ, Yan D, Xu Y. Nylon 10, 10-MMT nanocomposites made by intercalating polymerisation. Polymer Engineering and Science. 2003;**43**(1):204-213

[38] Hofmann B, Dietrich C, Thomman R, Friedrich C, Mulhaupt R. Morphology and rheology of PS clay

nanocomposites. Macromolecular Rapid Communications. 2000;**21**:57-61

[39] Huang SH, Paeng SW, Kim JY, Huh W, Lee SW. Synthesis of allylester resin tethered to layered silicates by in-situ polymerization and its nanocomposites. Polymer Bulletin. 2003;**49**:329-335

[40] Wang S, Hu Y, Song L, Wang Z, Chen Z, Fan W. Preparation and thermal props of ABS-MMT nanocomposites. Polymer Degradation and Stability. 2002;**77**:423-426

[41] Chaudhary DS, Prasad R, Gupta RK, Bhattacharya SN. Clay intercalation & influence on crystallinity of EVA based clay nanocomposites. Thermochimica Acta. 2005;**433**:187-195

[42] Li B, Hu Y, Zhang R, Chen Z, Fan W. Preparation of poly (vinyl alcohol)/layered double hydroxide nanocomposite. Materials Research Bulletin. 2003;**32**:1567-1572

[43] Strawhecker KE, Manias E. Structure and properties of poly (vinyl alcohol)/Na^+ MMT nanocomposites. Chemistry of Materials. 2000;**12**:2943-2949

[44] Tseng C, Wu J, Chang F. Crystallisation behavior of syndiotactic PS nanocomposites for melt and cold crystallisation. Journal of Applied Polymer Science. 2002;**86**:2492-2501

[45] Zhang G, Shichi T, Takagi K. PET clay hybrids with improved tensile strengths. Materials Letters. 2003;**57**:1858-1862

[46] Ko MB. Effects of acrylonitrile content on the properties of clay dispersed poly (styrene-co-acrylonitrile) copolymer nanocomposite. Polymer Bulletin. 2000;**45**:183-190

[47] Brahim C, Ammari F, Meganem F. Involvement of hybrid materials based on natural clay and poly (vinylchloride) into extraction of metal cations. Polimery. 2017;**62**(3):187-192

[48] Mallick PK, Zhou Y. Yield & fatigue behavior of PP & PAM-6 nanocomposites. Journal of Materials Science. 2003;**38**:3180-3190

[49] Gu A, Chang FC. Preparation of PI/clay hybrid films. Journal of Applied Polymer Science. 2001;**79**:289-294

[50] Xu R, Manias E, Snyder AJ, Runt J. Biomedical PU layered silicate nanocomposites. Macromolecules. 2001;**34**:337-339

[51] Hu Y, Song L, Xu J, Yang L, Chen Z, Fen W. Synthesis of PU/clay intercalated nanocomposites. Colloid Polymer. 2001;**279**:819-822

[52] Chang JH, Uk AY, Ryu SC, Giannelis EP. Synthesis of PBT/clay nanocomposites thro' in-situ interlayer polymerization. Polymer Bulletin. 2003;**51**:69-75

[53] Chang JH, Park DK. Various organo clays based nanocomposites of PET. Polymer Bulletin. 2001;**47**:191-197

[54] Ianchis R et al. Synthesis and properties of new epoxy-organo layered silicate nanocomposites. Applied Clay Science. 2011;**52**:96-103

[55] Silva AA et al. Nanostructure and dynamic mechanical properties of silane functionalized montmorillonite/epoxy nanocomposites. Applied Clay Science. 2011;**54**:151-158

[56] Huskic M, Zigon M, Ivanokovic M. Comparison of the properties of clay polymer modified by silane and by quarternary ammonium salts. Applied Clay Science. 2013;**85**:109-115

[57] Huang G-B, Ge CH, He B-J. Preparation, characterization and properties of amino functionalized montmorillonite and composite layer-by-layer assembly with inorganic

nanosheets. Applied Surface Science. 2011;**257**:7123-7128

[58] Kim Y et al. New sulfonic acid moiety grafted on montmorillonite as filler of organic-inorganic composite membrane for non-humidified proton exchange membrane fuel cells. Journal of Power Sources. 2010;**195**:4653-4659

[59] Romanzini D et al. Effect of clay silylation on curing and mechanical and thermal properties of unsaturated polyester/montmorillonite nanocomposites. Journal of Physics and Chemistry of Solids. 2015;**87**:9-15

[60] Mishra AK et al. Characterization of surface modified montmorillonite nanocomposites. Ceramics International. 2012;**38**:929-934

[61] Chavarria F, Paul DR. Comparison of nanocomposites based on nylon 6 and nylon 66. Polymer. 2004;**45**:8501-8515

[62] Kawasumi M, Hasegawa N, Kato M, Usuki A, Okada A. Preparation and mechanical properties of PP clay hybrids. Macromolecules. 1997;**30**:6333-6338

[63] Gultek A, Galip Icduygu M, Seckin T. Preparation and characterization of PBI clay hybrid materials. Materials Science and Engineering. 2004;**B107**:166-171

[64] Rehab A, Salahuddin N. Nanocomposite materials based on PU intercalated into montmorillonite clay. Materials Science and Engineering. 2005;**A399**:368-376

[65] Yao KJ, Song M, Hourston DJ, Luo DZ. Polymer/layered clay nanocomposites: 2 polyurethane nanocomposites. Polymer. 2002;**43**:1017-1020

[66] Hu Y, Song L, Xu J, Yang L, Chen Z, Fan W. Synthesis of polyurethane/clay intercalated nanocomposites. Colloid & Polymer Science. 2001;**279**:819-822

[67] Gu A, Chang FC. Novel preparation of PI/clay hybrid films with low co-efficient of thermal expansion. Journal of Applied Polymer Science. 2001;**79**:289-294

[68] Kornmann X, Lindberg H, Berglund LA. Synthesis of epoxy-clay nanocomposites: Influence of the nature of the clay on structure. Polymer. 2001;**42**:1303-1310

[69] Van Den Oever MJA, Elbersen HW, Keijsers ERP, Gosselink RJA, De Klerkengels B. Switchgrass as a reinforcing fiber in PP composites. Journal of Materials Science. 2003;**38**:3697-3707

[70] Mallick PK, Zhan YX. Yield & fatigue behavior of PP & PAM-6 nanocomposites. Journal of Materials Science. 2003;**38**:3183-3190

[71] Garcia Lopez D, Picazo O, Merino JC, Pastor JM. PP clay nanocomposites. European Polymer Journal. 2003;**39**:945-950

[72] Weibing X, Liang G, Zhai H, Tang S, Hang G, Pang WP. Preparation & crystallization behavior of PP/PP-g-MAH/org-MMT nanocomposites. European Polymer Journal. 2003;**39**:1467-1474

[73] Li J, Zhou C, Gang W, Wei Y, Ying T. Preparation and linear rheological behavior of PP/MMT nanocomposites. Polymer Composites. 2003;**24**(3):323-331

[74] Collar EP, Laguna O, Areso S, Garcia-Martinez JM. Succinyl fluorescein grafted atacted PP as an interface modifier in PP/talc composites. European Polymer Journal. 2003;**39**:157-163

[75] Kato M, Okamato H, Hasegawa N, Tsukigase A, Usuki A. Preparation and properties of PE clay hybrids. Polymer Engineering & Science. 2003;**43**(6):1312-1316

[76] Kuo S, Huang W, Huang S, Kao HS, Chang FC. Synthesis & characterization

of in-situ blended metallocene PE clay nanocomposites. Polymer. 2003;**44**:7709-7719

[77] Shin SYA, Simon LC, Soares JBP, Scholz G. PE clay hybrid nanocomposites. Polymer. 2003;**44**:5317-5321

[78] Zheng H, Zhang Y, Peng Z, Zhang Y. Influence of clay modification on the structure and mechanical properties of EPDM MMT nanocomposites. Polymer Testing. 2004;**23**:217-223

[79] Li J, Zhou C, Gang W. Study on nonisothermal crystallization of maleic anhydride grafted PP/MMT nanocomposite. Polymer Testing. 2003;**22**:217-223

[80] Zheng X, Jiang DD, Wilkie CA. Polystyrene nanocomposites based on an oligomerically modified clay containing maleic anhydride. Polymer Degradation and Stability. 2006;**91**:103-113

[81] Wang KH, Choi MH, Koo CM, Choi YS, Chung IJ. Synthesis & characterisation of maleated PE/ clay nanocomposites. Polymer. 2001;**42**:9819-9826

[82] Chao Chen Em M et al. PE/clay hybrid materials: A simple method to modulate the optical properties. Polimeros. 2016;**26**(1):38-43

[83] Teichler A et al. Inkjet printing of chemically tailored light emitting polymers. European Polymer Journal. 2013;**49**(8):2186-2195

[84] Manickandan Nair KC, Thomas S. Interface modification on the mechanical properties of PS-sisal composites. Polymer Composites. 2003;**24**(3):332-343

[85] Saxena M, Gauri VS. Studies on bamboo polymer composites. Polymer Composites. 2003;**24**(3):428-436

[86] Papargyri SA, Cooke RG, Papargyris DA, Botis A, Papapolmeron G, Papargyris AD. Kaolin clay matrix composites. British Ceramic Transactions. 2003;**102**(5):193-203

[87] Thwe MM, Liao K. Environmental effects on bamboo-glass/PP hybrid composites. Journal of Materials Science. 2003;**38**:363-376

[88] De Carvallo AJF, Curvello AAS, Agnelli JAM. Composites of thermoplastic starch and kaolin. Carbohydrate Polymers. 2001;**45**:189-194

[89] Okada K, Mitsunaga T, Nagase Y. Properties and particle dispersion of biodegradable resin/clay nanocomposites. Korea-Australia Rheology Journal. 2003;**15**(1):43-50

[90] Hussain M, Simon GP. Phosphorous-clay polymer nanocomposites for fire performance. Journal of Materials Science Letters. 2003;**22**:1471-1475

[91] Rodlert M, Christopher JG, Garamszegi L, Leterrier Y, Grunbauer HJM, Manson J-AE. Hyperbranched polymer/ MMT clay nanocomposites. Polymer. 2004;**45**:949-960

[92] Ratna D, Becker O, Krishnamurthy R, Simon GP, Varley RJ. Nanocomposites based on a combination of epoxy resin, hyperbranched epoxy, and a layered silicate. Polymer. 2003;**44**:7449-7457

[93] Nassar EJ, Neri CR, Cale EPS, Serra OA. Functionalised silica synthesized by sol-gel process. Journal of Non-Crystalline Solids. 1999;**247**:124-128

[94] Kickelbick G. Concepts for the incorporation of inorganic building blocks into organic polymers on a nanoscale. Progress in Polymer Science. 2003;**28**:83-114

Chapter 3

Structural Changes in Vermiculites Induced by Temperature, Pressure, Irradiation, and Chemical Treatments

Celia Marcos

Abstract

Depending on the treatment, the crystallinity increase of vermiculite may be accompanied by the enhancement of the majority starting phase, and the crystallinity loss may be accompanied by the appearance or disappearance of interstratified phases. Starting vermiculites with high K^+ content in the interlayer have more interstratified phases and lower water content and are less crystalline. The crystallinity loss of vermiculite and therefore the structural disorder increase are caused by the structural water loss. On the contrary, the crystallinity increase is produced by water gain. The vermiculite transformation by structural water loss occurs with temperature increase, vacuum, irradiation with microwaves or ultraviolet, and alcohol or acidic treatment. On the contrary, the transformation by water gain occurs in vermiculites treated with hydrogen peroxide and in those subjected to ionic metal exchange. These treatments provide evaluable information on the relationship between the structure of vermiculites and their industrial applications. The changes suffered by vermiculites due to the treatments applied could give light to ambiguities about their geological origin and hydrothermal and/or supergene processes.

Keywords: vermiculite, structural changes, physical and chemical treatments

1. Introduction

Vermiculite is a mineral that belongs to the phyllosilicate subclass of the silicate class. It has an appearance similar to micas at macroscopic level (**Figure 1**), with varied colors (green, yellow, to brown), leafy habit, hardness about 2, and a density between 2.4 and 2.7 g/cm^3.

Its structure (**Figure 2**) corresponds to that of the 2:1 group [1], which is composed of two T-O-T layers joined by an interlayer. The T-O-T layer is composed of an octahedral (O) sheet of Mg^{2+}, located between two tetrahedral sheets (T) of Si^{4+}. The interlayer is formed by an octahedral sheet of Mg^{2+} bound to oxygens or OH^- groups. In addition, it contains water.

In vermiculite, isomorphic substitutions, especially in the tetrahedral sheets of Si^{4+} to Al^{3+}, are very common. As a consequence of the positive charge difference, compensation occurs with cations in the interlayer space, mainly Mg^{2+}, as mentioned before.

Figure 1.
Vermiculite appearance in hand sample.

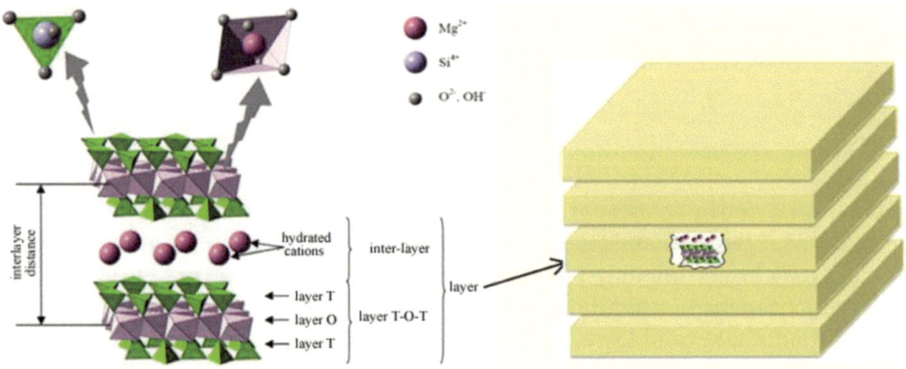

Figure 2.
Vermiculite structure.

This structure, with spatial group C2/c, is generally disordered [2], that is, it shows stacking defects that alter the regular alternation of the layers parallel to crystallographic axis b (**Figure 3**).

Due to the presence of water and OH^- groups, vermiculite can undergo hydration-dehydration processes that depend on various factors such as temperature, pressure, particle size, relative humidity, and chemical composition [1, 3–10].

The hydration state of vermiculites is defined by the number of layers of water in the interlayer space, with phases having zero, one, and two water layers. These phases were named by [11] as 0-WLHS (state of hydration with 0 water layers), 1-WLHS (state of hydration with 1 water layer), and 2-WLHS (state of hydration with 2 water layers), respectively. As an example, for Mg-vermiculite the basal distances are 9.02 Å for 0-WLHS, 11.50 Å for 1-WLHS, and 14.40 Å for 2-WLHS [9–12].

The chemical formula of vermiculite is $X_4(Y_{2-3})O_{10}(OH)_2M,nH_2O$, where X represents the tetrahedral positions (Si^{4+} y Al^{3+}), Y the octahedral positions (Mg^{2+}, Fe^{2+}, Fe^{3+}, Cr^{3+}, Ti^{4+}, etc.), and M the cations located in the interlayer space (Mg^{2+}, Ca^{2+}, K^+, Na^+, etc.) to compensate the charges, as a consequence of the isomorphic substitutions.

In addition to the described mineral that corresponds to vermiculite in the strict sense, there are the so-called commercial vermiculites. These vermiculites consist of various interstratified of mica/vermiculite, vermiculite with different states of hydration, mixtures of mica and vermiculite, etc. The distribution of the different phases would be mosaic-type (**Figure 4**).

The main characteristic of commercial vermiculites is their exfoliation and expansion capacity when the vermiculite is abruptly heated, and that occurs due to the loss of water molecules located between the silicate sheets (**Figure 5**).

Figure 3.
Disordered layers in the vermiculite structure.

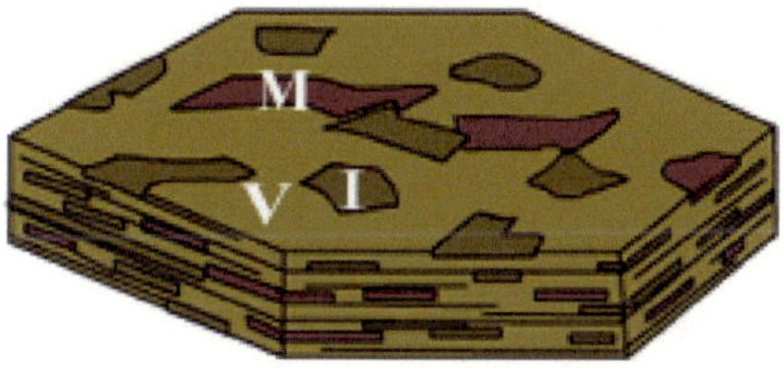

Figure 4.
Mosaic distribution of different phases in a commercial vermiculite (modified from Hillier et al. [15], with permission).

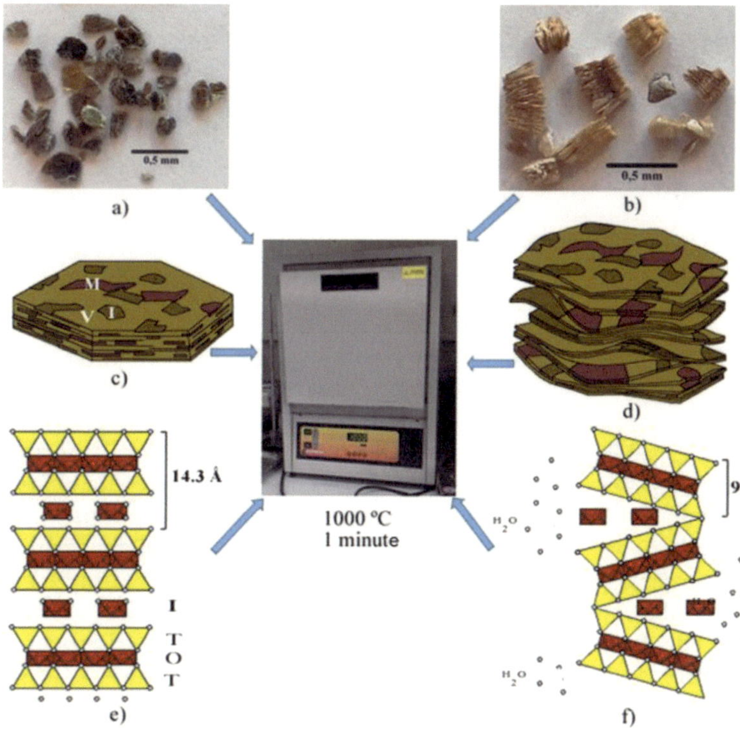

Figure 5.
Exfoliation mechanism scheme of a commercial vermiculite when heated at 1000°C for 1 minute; (a) commercial vermiculite; (b) exfoliated commercial vermiculite; (c) scheme of the arrangement in domains of the different intergrowth phases in a particle of a commercial vermiculite; (d) diagram of the exfoliated particle; (e) structure of an ideal vermiculite; (f) structure of the exfoliated ideal vermiculite. Note: Schemes (c) and (d) modified from Hillier et al. [15] with permission.

Authors such as [13] and later [14] found that the greatest exfoliation is achieved in the case of regular mica-vermiculite interstratified. Couderc and Douillet [14] associated this fact with the collision, during the "thermal shock," of the water molecules of the vermiculite sheets with the mica sheets, producing a greater separation between them; Hillier [15] related exfoliation with the mosaic distribution of the different mineral phases within the vermiculite particles. Lateral phase boundaries between vermiculite and other phases (mica, or vermiculite, and chlorite) would prevent vapor from escaping from a particle, resulting in exfoliation when the pressure exceeds the bonding forces that hold the layers together. This type of thermal exfoliation is the oldest and the one that is still used today mostly in the industry.

Vermiculites can be modified by changes in temperature and pressure, chemical treatments, and irradiation, causing physical and structural changes in the mineral [16–24].

One of the most notable physical changes is exfoliation and expansion, which are influenced by factors such as water content, type of cations of the interlayer, and interstratifications of vermiculite [10, 17, 25].

Unmodified and modified commercial vermiculites are characterized by their industrial and technological applications [26–35]. These applications are a function of its physical and chemical properties and the treatment it has undergone, for example, thermal and acoustic insulation, adsorbent of substances, refractories, fire protection, support for hydroponic crops, light concretes, etc. The synthesis of advanced materials such as new glasses of great technological interest constitutes an example of the uses based on chemical applications. Numerous studies on the intercalation of polar organic molecules by clay minerals have been carried out. In addition to water, inorganic or organic substances can be adsorbed in the expandable interlayer space [36, 37].

The objective of this research has been to show the structural changes in commercial vermiculites induced by temperature, pressure, irradiation and chemical treatments, the relationship between these treatments, and crystallinity and the possible causes.

2. Methodology

2.1 Materials

The investigated vermiculite samples come from Catalão (Goiás, Brazil), Paulistana (Piauí, Brazil), China, Libby (Montana, USA), Benahavis (Málaga, Spain), and Sta. Olalla (Huelva, Spain) to compare. Vermiculite from Catalão (hereinafter Goiás) is associated with an ultramafic complex; Paulistana's vermiculite (hereinafter Piauí) is found in a hybrid basic rock, probably a lamprophyre [38]. The origin of China's vermiculites is unknown. The origin and mineralogy of the vermiculite of Sta. Olalla have been extensively studied [39–42]. This vermiculite is formed from phlogopite as a result of the alteration of pyroxenites. Vermiculite from Benahavis occurs in elongated veins, and the host rock is mainly serpentine [43, 44] and can be considered formed by alteration of phlogopite [45].

The weight percent of element oxides of the vermiculites considered in this chapter [10, 46] is in **Table 1** and their water content (%) in **Table 2**.

2.2 Vermiculite treatments

2.2.1 Heat treatment

For the experiments with heat [10, 23], two diffractometers were used: Seifert XRD 3000 diffractometer (Scientific-Technical Services of the University of

Sample	Sta. Olalla[1]	Benahavis[1]	Piauí[1]	Goiás[1]	China W[1]	China G[1]	Palabora[1]	Libby[2]
SiO_2	35.9	37.0	39.9	40.7	43.2	35.6	41.1	38.7
TiO_2	0.3	2.5	1.1	0.8	1.0	1.2	1.2	1.2
Al_2O_3	15.8	14.1	9.3	11.5	11.9	11.0	10.0	13.0
Cr_2O_3	0.0	0.0	0.1	0.0	0.2	0.4	0.0	1.0
FeO	3.3	7.6	6.7	9.6	4.3	4.6	7.9	8.6
MnO	0.1	0.1	0.0	0.1	0.0	0.0	0.0	0.1
MgO	24.1	21.9	25.5	18.0	24.3	21.8	23.3	20.6
CaO	0.3	0.1	0.2	0.0	0.4	0.9	0.2	0.0
Na_2O	0.1	0.1	0.0	0.1	0.7	3.5	0.1	0.3
NiO	0.0	0.1	0.0	0.0	0.0	0.1	0.1	0.0
K_2O	0.0	0.0	3.5	1.1	7.5	5.6	6.0	9.7

[1]*Marcos et al. [9].*
[2]*Marcos and Rodríguez [46].*

Table 1.
Weight percent of element oxides.

	Water content (%)
Sta. Olalla[1]	25.6
Piauí[1]	14.0
Goiás[1]	13.6
China O[1]	17.6
China G[1]	12.3
Palabora[1]	13.4
Libby[2]	10.3

[1]*Marcos et al. [9].*
[2]*Marcos and Rodríguez [46].*

Table 2.
Water content (%).

Oviedo) at 30 mA and 40 kV; Cu-K_α radiation, $\lambda = 1.5418$ Å; 2θ range of 3–20°; 2° scans of 0.02° per step; and a count time of 20 s per step and Bruker AXS diffractometer (Plasma Physics Laboratory, National University, Manizales headquarters) at 30 mA and 40 kV (Cu-Kα radiation; $\lambda = 1.5418$ Å), 2θ range of 3–40°, 2θ scans of 0.1°, and a count time of 20 s per step.

With an increase of T, the behavior of the vermiculites was different depending on the composition of the vermiculite and the type of heating (ex situ or in situ).

The result with heating ex situ at 1000°C for 1 minute was an expanded and exfoliated light product composed of enstatite in the purest vermiculites (with Mg^{2+} or Mg^{2+} and K^+ in the interlayer) Sta. Olalla and Benahavis and mica and enstatite in the commercial vermiculites (with K^+ and/or Na^+ and/or Ca^{2+} in the interlayer). The exfoliation of the latter was much greater than that of the former.

The diffraction patterns made with temperature increase (40–140°C) in situ using the Seifert XRD 3000 equipment, and a powder sample showed the coexistence of the 2-WLHS phases with two layers and one layer of water, and the 1-WLHS phase was revealed. In the patterns made with the Bruker AXS diffractometer, the last phase did not appear, and it was observed that the structure practically collapsed at 100°C, phase 1-WLHS reappearing as temperature increases.

With gradual increase in T in situ, dehydration of the vermiculites could be observed until practically reaching collapse, although the behavior was different depending on whether the samples contained Mg^{2+} or Mg^{2+} and K^+ in the interlayer or K^+ and/or Na^+ and/or Ca^{2+}. In the former dehydration appears to be restricted to 1-WLHS, and dehydroxylation begins at lower temperatures and is faster than in the latter, in which already dehydrated vermiculite coexists with a similar structure to mica.

2.2.2 Pressure treatment

The experiments at atmospheric pressure $P = 1.4{\cdot}10^{-2}$ mbar and $P = 2.4{\cdot}10^{-4}$ mbar were carried out on both powder samples and exfoliation flakes of three vermiculites from Sta. Olalla (Huelva, Spain), Paulistana (Piauí, Brazil), and Western China [9]. A Seifert XRD 3000 diffractometer from the Scientific-Technical Services of the University of Oviedo was used. The conditions of use were 30 mA and 40 kV (Cu-Kα radiation, $\lambda = 1.5418$ Å), range 2θ between 2 and 70°2θ, speed 0.02°2θ/20 s. Two commercial Leybold pumps (Trivac D 2.5 E (up to 10–2 mbar) and Turbovac TMP 50 (up to 10–4 mbar)) were used for the vacuum experiments.

The effect of vacuum, like that of temperature, causes dehydration of vermiculite but with a different evolution through the different states of hydration. In fact, under vacuum, the process appears to inhibit itself to a state of hydration with one layer of water (1-WLHS). The role of T is inhibited against that of pressure. The Sta. Olalla vermiculite gave rise to the formation of three different interstratified phases: two phases characterized by an interstratification with interplanar distances, $d = 11.5$–13.8 Å and $d = 9.6$–11.5 Å, respectively, and another phase with $d = 13.8$ Å.

Under vacuum, $P = 1.4{\cdot}10^{-2}$ mbar, in the Sta. Olalla, vermiculite phase 2-WLHS was observed with two layers of water coexisting with phase 2-WLHS but with a layer of water. In Piauí vermiculite, the evolution of the most characteristic reflection was more remarkable since it disappears, appearing in phase 1-WLHS.

The effect was faster with powder samples, and phase 2-WLHS quickly transforms to phase 1-WLHS.

2.2.3 Irradiation treatment

2.2.3.1 Microwave irradiation

The transformations undergone by the vermiculites subjected to microwave irradiation (at 800 W and exposure times from 10 to 20 s) were characterized by X-ray diffraction [24]. A PHILIPS X'PERT PRO X-ray diffractometer was used, at 40 mA and 45 kV, Cu-Kα radiation ($\lambda = 1.5418$ Å), range of 3–70°2θ, steps of 0.02°, and a count time 1 s per step.

Microwave irradiation of vermiculite samples caused much less water loss than they do when subjected to sudden high-temperature heating; it also caused exfoliation of the material. From a structural point of view, the X-ray diffraction patterns of the vermiculites of Sta. Olalla, China, and Libby showed loss of crystallinity and disorder.

2.2.3.2 Ultraviolet irradiation

The changes induced by ultraviolet (UV), short wave (254 nm), and long (356 nm) radiation at different times (1 hour, 1 day, and 1 week) in vermiculites from Sta. Olalla, Libby, and China were studied by using X-ray diffraction [46]. A PHILIPS X'PERT PRO X-ray diffractometer was used, at 40 mA and 45 kV,

Cu-Kα radiation (λ = 1.5418 Å), range of 3–10°2θ, steps of 0.007°, and a count time 1 s per step. Crystallite size and structural deformation were evaluated using PANalytical software (X'Pert Plus).

In the powder samples of Sta. Olalla, Libby, and China, a decrease in the intensity of the most characteristic reflections was observed as well as a decrease in the size of the crystallite and an increase in deformation. The results were more significant for powder vermiculite from China than for Sta. Olalla and Libby vermiculites, probably due to the coexistence of different hydration states, interstratifications, and superstructures in the initial vermiculite in China. The water loss in these long and short UV irradiated samples for 168 hours was 4 and 4.5% for the Sta. Olalla sample, 0.9 and 1% for the Chinese sample, and 6.4 and 7% for the Libby sample. In the exfoliation flake samples, an increase in the intensity of the most characteristic reflection is observed, in addition to a larger crystallite size and a lower percentage of deformation, so that the crystallinity increased.

2.2.4 Chemical treatment

By **ion exchange of metals** (Ni^{2+}, Fe^{2+}, Fe^{3+}) for the Mg^{2+} of the interlayer in the vermiculite-Mg of Sta. Olalla (Huelva, Spain).

In the case of nickel [35], an aqueous solution of Ni^{2+} acetate was used, and in the case of iron [47], $FeCl_2$ and $FeCl_3$ solutions were used. The crystal structures of the nickel and iron vermiculites were refined using the DIFFaX+ program [48] (and a later version). The characterization was carried out with X-ray diffraction by transmission using an InEL XRD RG3000 cobalt tube vertical diffractometer (λ = 1.7890 Å) (40 kV, 35 mA) and an InEL CPS 120 detector, in the 2θ range of 4–70° (step 0.03°, total acquisition time of 5400 s). Refinement confirmed the similarity of the Ni^{2+}-, Fe^{2+}-, and Fe^{3+}-vermiculites with the Mg-vermiculite. In the Fe^{2+}-vermiculite reflections corresponding to the akaganeite (β-FeOOH) (JCPDS card 01–075-1594), a phase that was corroborated by Mössbauer spectroscopy (16%) was observed, a technique that also allowed showing the presence of said phase (3%) in the Fe^{3+}-vermiculite. Only vermiculite was detected in the Ni^{2+}-vermiculites obtained from the starting vermiculite, while in the one obtained from the homoionized starting vermiculite (with brucite -Mg $(OH)^{2-}$), a brucite phase with magnesium and nickel was also detected. In this refinement it was found that Ni^{2+} and Fe^{2+} also enter the octahedral layer, although there was no evidence for Fe^{3+}.

2.2.4.1 Acid activation with HNO_3

Nitric acid treatment at 4 and 8 M was used at room temperature and different treatment times on the purest vermiculite from Sta. Olalla and two commercial vermiculites from Goiás and China, respectively [49]. To quantify potential loss of mass or water, 1 mL of each sample was weighed pre- and post-acidic treatment, and their volume post-treatment was measured. To identify any structural change, X-ray diffraction patterns were taken with a PANalytical X'pertPro diffractometer using 40 mA and 45 kV (Cu-Kα radiation; λ = 1.5418 Å), 2θ scans 5–35o, 2θ step scans of 0.007°, and a counting time of 1 s per step. TEM high-resolution microscope with a resolution of 1.9 Å between points and 1.0 Å between lines was used to obtain TEM and selected area electron diffraction (SAED) micrographs with its accompanied CCD camera (Gatan).

Vermiculites treated with acid suffered (a) slight delamination and color (**Figure 6**); (b) weight loss (**Table 3**) due to the mass and water loss; (c) inhomogeneous cation leaching, probably achieved to transfer iron from those octahedral sheets to clusters deposited on vermiculite layers (**Figure 7**) [50–53] (and in the samples with

Figure 6.
Color of treated vermiculite from Sta. Olalla.

Samples	HNO_3 molarity	Time	Weight loss (%)
Sta. Olalla	4	1	70
		7	75
	8	1	72
		7	70
Goiás	4	1	12
		7	14
	8	1	47
		7	47
China	4	1	22
		7	27
	8	1	53
		7	52

Marcos et al. [49].

Table 3.
Weight loss (%) in the treated samples.

high iron content, this element would have prevented further leaching of cations but not water loss); and (d) structural transformation that resulted in the formation of lamellar products with low crystallinity and order, composed by amorphous silica and other phases whose identity and percentage varied depending on the vermiculite type.

2.2.5 Thermal and chemical treatment

2.2.5.1 Thermal treatment and subsequent reaction with metal ions solution (Cr^{3+} y Ni^{2+})

For the experiments with Cr^{3+} [54] and Cr^{6+} [55], commercial vermiculite from China thermo-exfoliated at 900°C for 1 minute was used. With Ni^{2+}, commercial from Piauí (Brazil) and China, vermiculites thermo-exfoliated at 1000°C for 1 minute were used [35].

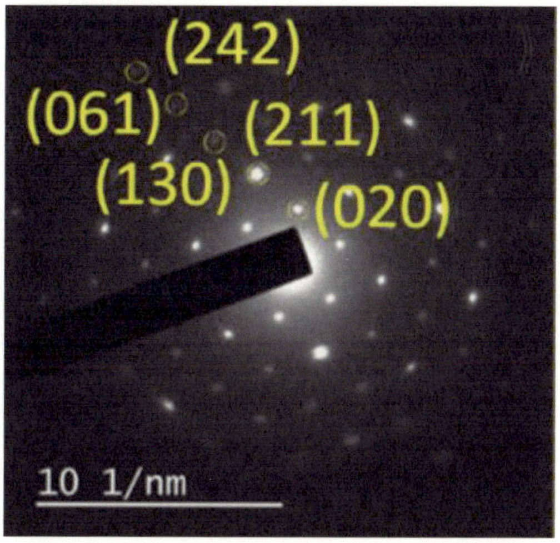

Figure 7.
TEM and SAED micrograph of polycrystalline goethite in Goiás vermiculite .

Vermiculite characterization after Cr^{3+} adsorption was performed with X-ray diffraction using a Bruker AXS D8 Advance diffractometer with an Anton Paar HTK1200 oven at room temperature and 900°C for 1 minute. The equipment conditions were 30 mA and 40 kV, Cu-Kα radiation (λ = 1.5418 Å), range of 3–40°2θ, 0.1° steps, and a count time of 20 s per step.

The X-ray spectra of Chinese vermiculite after having been in contact with a solution of synthetic seawater and dissolved Cr^{3+} with concentrations of 0.75 and 2.0 ppm suggested that the mica-like phase of the exfoliated vermiculite would have been transformed back into a vermiculite-like structure, similar to that of the original sample. In contrast, in the X-ray spectrum of Chinese vermiculite after having been in contact with distilled water solution and dissolved Cr^{6+} with concentrations of 1 ppm (**Figure 8**), no change is observed.

2.2.5.2 Hydrogen peroxide

Several experiments with vermiculite samples from Sta. Olalla (Huelva, Spain), Libby (Montana, USA), and Goiás (Brazil), in both powder and flakes forms, were carried out [56]: (a) treatment with 30% and 50% hydrogen peroxide solution, at different times for each sample; (b) irradiation with microwaves for 20 s of Libby and Goiás samples; and (c) treatment with 30% and 50% hydrogen peroxide solution in the microwave oven for 20 s of vermiculite samples from Goiás. X-ray diffraction was used to identify structural changes. For the powder samples, a PHILIPS X'PERT PRO diffractometer was used at 40 mA and 45 kV, Cu-K_α radiation (λ = 1.5418 Å), range of 3–70°2θ, steps of 0.02°, and a count time 1 s per step. For the flake samples, a Seifert XDR 3000 T/T diffractometer was used at 30 mA and 40 kV, Cu-K_α radiation (λ = 1.5418 Å), range of 2–10°2θ, steps of 0.02°, and a time of 20 s count per step. The results showed that the vermiculites hardly underwent changes at the structure level, despite the change in appearance and textural (color, exfoliation, corrugation, undulations) (**Figure 6**). The three vermiculites—Sta. Olalla, Goiás, and Libby—showed a slight increase in the intensity of the main reflection. The change was slower in powder form than in flake samples, at least in Libby vermiculite.

Alcohol (methanol, ethanol, propanol, butanol).

The structural changes of commercial vermiculites treated with alcohol and alcohol for 1 month and subsequent microwave irradiation for 20 seconds were

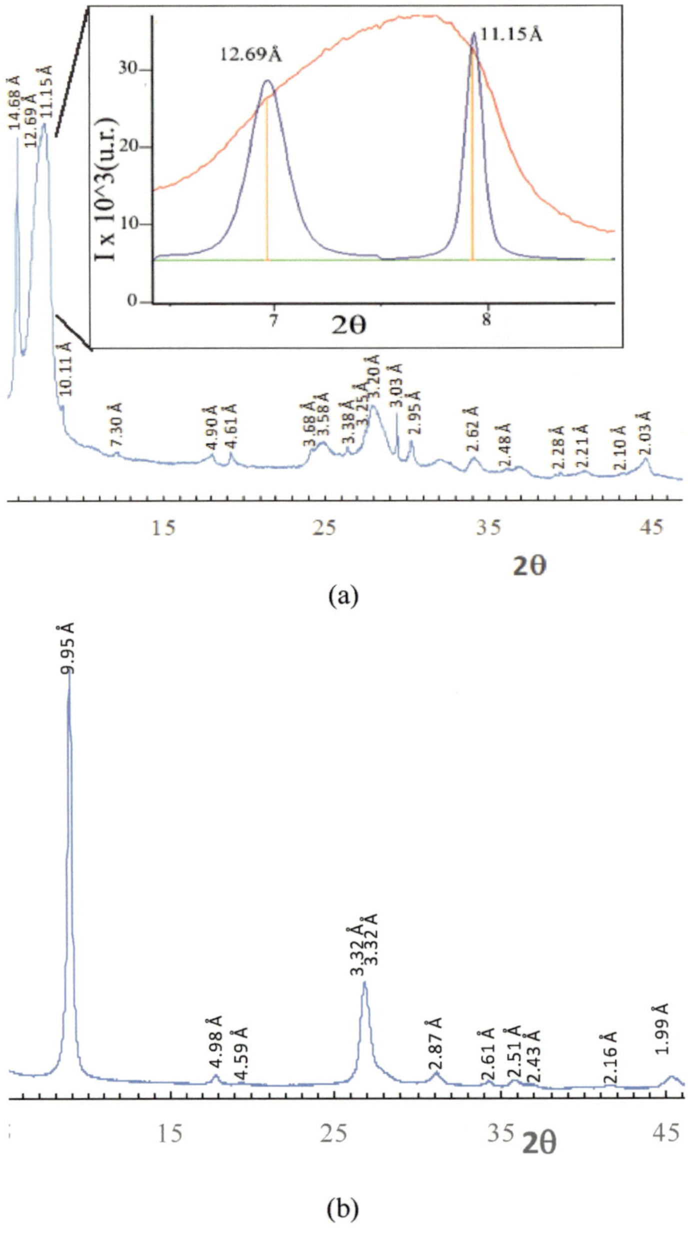

Figure 8.
X-ray diffraction of the starting Chinese vermiculite and abruptly heated at 900°C for 1 minute (a) and after having been in contact with 1 ppm Cr^{6+} in distilled water (b).

analyzed using X-ray diffraction [57, 58]. The equipment used was a PHILIPS X'PERT PRO diffractometer, at 40 mA and 45 kV, Cu-Kα radiation (λ = 1.5418 Å), range of 3–12°2θ, steps of 0.007°, and a count time of 1 s per step. Changes in the intensity and position of the basal reflections were used to indicate changes in the structural order and in the hydration states. In Sta. Olalla vermiculite, the intensity of the most characteristic reflection decreased with all the alcohols and times investigated. In the Chinese and Libby vermiculites, the intensity of the most characteristic reflections with methanol and ethanol also decreased, while with propanol and butanol, as a function of time, there was an increase in intensity and optimization of the profile of the aforementioned reflections and incipient appearance of phases with different hydration states (in Chinese vermiculite).

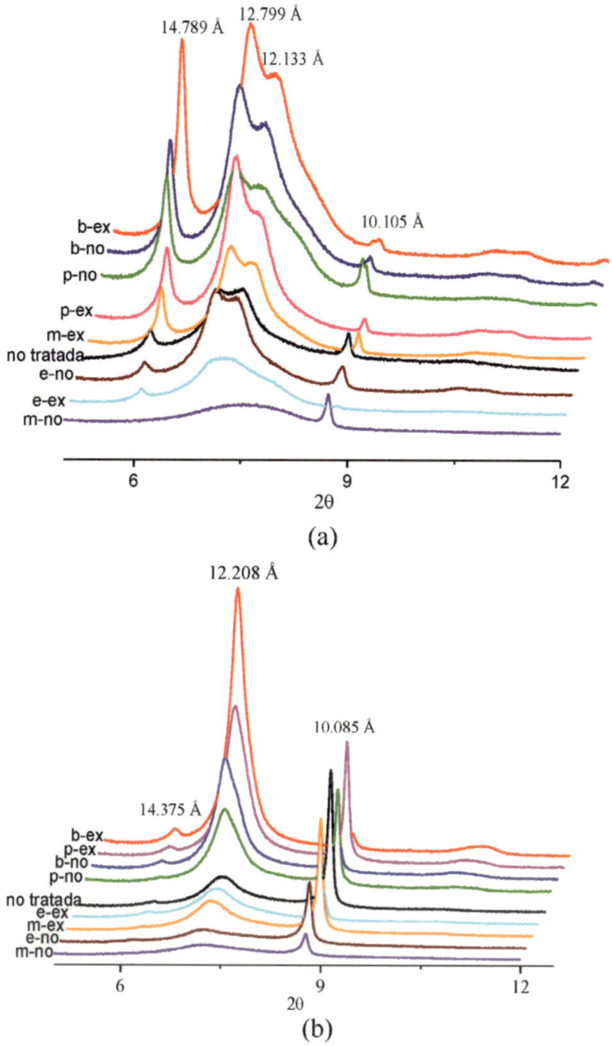

Figure 9.
X-ray diffraction of the samples from China (a) and Libby (b) treated with alcohol for 1 month and subsequent microwave irradiation for 20 seconds (Marcos et al. [59]). Note: m = methanol, e = ethanol, b = butanol, p = propanol, ex = exfoliated, no = no exfoliated.

The diffraction spectra for the vermiculite particles from China and Libby treated with alcohol for 1 month and subsequent microwave irradiation for 20 seconds, exfoliated and non-exfoliated, are shown in **Figure 9** [59]. In these vermiculites treated with butanol or propanol and subsequent microwave irradiation, it should be noted that the crystallinity and the order of phases 2- and 2-1-WLHS of the exfoliated and non-exfoliated particles improved in relation to the untreated ones, except in Chinese exfoliated particles after butanol treatment and subsequent microwave irradiation. The opposite occurred with methanol or ethanol treatment and subsequent microwave irradiation, except in the Chinese exfoliated particles after methanol treatment and subsequent microwave irradiation.

3. Discussion

The structural modifications of the investigated vermiculites treated thermally, with vacuum, or irradiation or chemically, consist of the phase transformation and the increase or loss of crystallinity and, therefore, the increase or decrease of the structural order. Depending on the treatment, the increase in crystallinity may be accompanied by the appearance of the majority starting phase, and the loss of crystallinity may be accompanied by the appearance or disappearance of interstratified phases.

Vermiculite responds to an increase of T by transforming its structure, which affects its applications. It is a dynamic process that depends on the composition, size, and shape of the particle, relative humidity, and process conditions (in situ or ex situ heating). With T increase, the transformations occur due to dehydration, greater in the purest vermiculites and less in the most micaceous ones. With abrupt ex situ heating, at 1000°C, with powder samples, the purest vermiculites are transformed into enstatite and the least pure into mica and enstatite; in both cases they appear expanded and exfoliated. With gradual heating in situ, at 1000°C and in flake samples, the structure practically collapses in the purest and the most micaceous vermiculites; although it fails to do so, its crystallinity is very low. Dehydration would occur by the escape of water in the form of steam. This escape would occur when the vapor pressure exceeds the bonding forces that hold the layers together, causing exfoliation and expansion [15]. The process occurs faster in powder samples than in flake samples.

The effect of the vacuum is similar to that of the temperature increase; the transformation also occurs by dehydration, although the process seems to be inhibited to a state of hydration with a layer of water (1-WLHS), without additional dehydration of the samples up to a state of hydration of zero layers of water (0-WLHS). The loss of water was less than with an increase in T. The process is slower than with an increase in temperature in situ since the pressure does not imply an increase in the activation energy as with the first. In addition, it was shown that the dehydration process occurs through different interstratified states in vermiculite. This result has been related to the content of Mg^{2+} cations in the interlayer, due to its affinity with water molecules. The purest vermiculite of Sta. Olalla showed the most complex dehydration process due to its higher magnesium content in the interlayer. Due to its affinity for water, the higher the content of the cation, the greater the difficulty in eliminating water molecules. When the temperature and the vacuum are acting simultaneously, the sample is dehydrated just after the vacuum is established, and the temperature has no additional effect. The process, as with temperature increase, occurs more quickly in powder samples than in sheet samples.

Microwave irradiation of vermiculite samples caused a loss of water much lower than what they suffer when subjected to sudden heating at high temperature. The said dehydration was slower than with vacuum or with sudden heating at 1000°C. As a consequence, the phase with d = 13.8 Å in Sta. Olalla, for example, could not be observed. The crystallinity loss and structural disorder is attributable to the water loss. There was no collapse of the structure or formation of new phases, probably because this loss was much lower than that produced at a temperature higher than that produced by microwave radiation. The expansion process began in the flake center and advanced toward the edges. The alignment and reorientation of the water dipoles with the applied field generated internal friction that caused the heating of the vermiculite and the vaporization of many of the water molecules. The explanation of steam escape and exfoliation would be the same as when the temperature rises.

The decrease in crystallinity and structural order in vermiculite powder samples irradiated with long and short UV is also attributable to the water loss. On the contrary, in crystalline samples the crystallinity and structural order increased. In this case there was surely rehydration by ambient humidity adsorption.

In the chemical treatment by ion exchange of Ni^{2+} or Fe^{2+} or Fe^{3+} metals by the Mg^{2+} of the interlayer in the Mg- vermiculite of Sta. Olalla, the decrease in the interplanar distance d_{002} has been interpreted as due to the interaction of the cation and a possible modification of the interactions between the exchanged cation and the TOT sheets, due to both the nature of the new cation and the variations in the quantity and distribution of the H_2O molecules that it induces. The formation of brucite in the homoionized starting vermiculite could be due to the homoionization process itself, consisting of introducing the vermiculite in a solution of $MgCl_2$ in order to eliminate possible impurities such as Na^+, Ca^{2+}, etc., coexisting with Mg^{2+}. In the vermiculite intercalated with nickel from this homoionized vermiculite, a brucitic phase with magnesium and nickel was also detected.

According to Ravichandran and Sivasankar [50], the reaction with nitric acid first caused the replacement of the exchangeable cations (Mg^{2+}, Ca^{2+}, K^+, Na^+) by protons, which will subsequently attacked the layers. Secondly, partial leaching of Al^{2+}, Mg^{2+}, and Fe^{2+} and Fe^{3+} from the tetrahedral and octahedral layers occurred. The silicon remained in the form of amorphous silica and quartz; this lasts in a very low percentage, which disappeared with the increase of the acid concentration in any of the samples. The Sta. Olalla sample suffered greater leaching of cations and water loss than the Goiás and China samples, whose high iron content would have prevented further leaching of other cations.

The heat treatment of the vermiculites and subsequent reaction with synthetic seawater solutions with Cr^{3+} and Ni^{2+} would have caused the reappearance of the starting vermiculite, by rehydration. Probably, its union with the water trapped in the thermo-exfoliated structure may have made this structure closer to the original, that is, the mica-like structure would have moved back to that of the vermiculite. In aqueous solution Cr^{3+} would be in the form $[Cr(H_2O)_6]^{3+}$ and Ni^{2+} as $[Ni(H_2O)_6]^{2+}$; in both cases its adsorption in thermally expanded vermiculites would be controlled by different cooperative mechanisms: (1) cation exchange and (2) surface complexation reactions [60–63]. It is important to note that a very low percentage of Ni^{2+} could have precipitated in the pores or on the surface of vermiculite such as magnesium and nickel hydroxide, according to the findings in unheated Ni^{2+}-, Fe^{2+}-, and Fe^{3+}-vermiculites [35, 47]. The cation exchange process would be favored by the fact that Cr^{3+} and Ni^{2+} have an ionic radius (0.69 and 0.78 Å, respectively), similar to that of Mg^{2+} (0.72 Å). This ion exchange process would have taken place considering that the product resulting from the heating of vermiculite would constitute a heterogeneous system formed by one or more disordered crystalline phases (mica type) with hydroxyl groups and some cations between layers of the original phase and molecules of water.

This transformation would have occurred due to the rehydration of the sample in the adsorption process, which should be directly related to the characteristics of the cation involved [64]. The transformation in both cases would be consistent with the investigations carried out by Derkowski and McCarty [65] on the rehydration of dehydroxylated smectite in an environment of low water vapor. The equilibrium occurs in the solid/liquid interface, where the available centers located on the surface could be exchanged with the species in the solution. This adsorption process would be the opposite of what happens to vermiculite when it expands to 900°C.

The transformation could have occurred due to two aspects: on the one hand, the ionic exchange of ions of the interlayer with Na^+ and other ions of the saline solution and on the other hand the lower force of attraction with the water of the interlayer which has Na^+ in relation to Mg^{2+} or Ca^{2+} [66]. In the case of no transformation because there was no hydration or dehydration, since there was no change in weight before and after the adsorption of the ion by expanded vermiculite, probably due to the characteristics of the cation involved [64].

In the case of the adsorption of Ni^{2+} by vermiculite, the behavior would have been similar to that of the adsorption of Cr^{3+} [54], although it would have to be confirmed experimentally.

The reaction with hydrogen peroxide showed textural rather than structural changes. The water content (**Table 4**) was practically the same in both the Goiás and Libby samples treated with H_2O_2 and untreated. The absence of hydration-dehydration is the cause of no phase transformation.

The leached cations, in greater quantity those of the interlayer (Na^+, K^+, etc.) than those of the tetrahedral and octahedral layers, would be replaced by the H^+ ions of the solution [21]. These ions gave rise to two effects: (1) increase of the pH of the solution and (2) corrosion of the vermiculite particles.

The structural changes of commercial vermiculites treated with alcohol, dehydration-hydration, and disorder-order would be related to the replacement of water by alcohol in a very low percentage, with weight loss and, depending on the type of vermiculite, appearance of phases with state of hydration of less layers of water. The structural changes of commercial vermiculites treated with alcohol and subsequent irradiation with microwaves consist of an increase or decrease in crystallinity and order. The results indicated dehydration-hydration and structural order-disorder that would be related to the entry of alcohol into vermiculites by water replacement, that is, by loss of water. The changes occurred in a manner similar to those produced with temperature and vacuum and were less pronounced for the purest vermiculite.

Structural changes of vermiculites induced by the above mentioned treatments provide evaluable information on the relationship between the structure of vermiculite and their industrial applications. In vermiculite applications as intumescent fire barriers [67–69] where exfoliation at low temperatures is required, the treatment type, in this case microwave irradiation, is more important than the structural change suffered by vermiculite. In the case of vacuum, not only the type of treatment but also the structural changes suffered by the vermiculite influence, since under pressure the vermiculite could act as a deposit mineral and host contaminating elements. Vermiculite irradiated with ultraviolet radiation could be used as material for optoelectronic devices because this radiation is less penetrating and easier and cheaper to obtain than gamma radiation [70]. Fe^{2+}- and Fe^{3+}-vermiculites maintain its paramagnetic character; and Ni^{2+}- vermiculite behaves as a two-dimensional spin-glass system in which planar ferro- and antiferromagnetic interactions compete, responsible for the complex magnetic behavior found. Ni^{2+}- vermiculites are interesting materials to study experimentally or

Samples	Treatment					
	—	Microwave irradiation	1000°C	H_2O_2 30%	H_2O_2 50%	
Sta. Olalla	25.6	24.9	4.9[a]			H_2O wt (%)
Goiás	12.1	12.2	5.6	12.1 12.7[b]	12.1 12.9[b]	
Libby	10.3	11.1	3.4	10.9	11.4	

[a]*Marcos et al. [10].*
[b]*Samples treated with alcohol and irradiated simultaneously with microwaves.*

Table 4.
Water content (%) obtained by thermogravimetry of untreated and treated samples of Sta. Olalla, Goiás, and Libby.

in simulations, with applications to physics, chemistry, materials science, and artificial neural networks in computing [71]. Alcohol treatment and subsequent microwave irradiation may be the procedure for obtaining purest vermiculite from a less pure sample. Nitric acid treatment of vermiculites with high iron content resulted in a lamellar products with high porosity, important in many applications such as low cost and efficient and sustainable adsorbent for dyes and metals [60, 72–74]. It is important to highlight how exfoliated vermiculites can remain unchanged depending on the valence of the adsorbed ion and the salinity and pH of the medium.

Consequently, the relationship between structural changes of vermiculite and the chemical and physical treatments could contribute to predicting the structural order–disorder of the vermiculite; the obtaining of purest vermiculite; the environmental fate of toxic metals, such as cesium (radioactive metal) in contaminated areas; and developing methods to extract these metals from contaminated soils or waters [75].

Further, the changes suffered by vermiculites due to the treatments applied could give light to ambiguities about their geological origin due to hydrothermal and/or supergene processes. However, most and possibly all macroscopic vermiculite and interstratifications of vermiculite and other phases (mica, chlorite) are believed to be of supergene origin [76, 77]. The changes suffered by vermiculites due to hydrogen peroxide treatment and ionic metal exchange, with water gain, could point to this origin, corroborating both the field and laboratory evidence in early times [76]. Regarding the treatments that involve water loss in vermiculites, it is not discarded that a more detailed study helps to reveal data related to the hydrothermal origin. Some aspects observed in the transformations caused by treatments with water loss could coincide with field observations [78, 79].

4. Conclusions

Starting vermiculites with high K^+ content in the interlayer have more interstratified phases and lower water content and are less crystalline.

The crystallinity loss and therefore the structural disorder increase are caused by the structural water loss. On the contrary, the crystallinity increase is produced by water gain.

The vermiculite transformation by structural water loss occurs with temperature increasing, vacuum, irradiation with microwaves or ultraviolet, and both alcohol and acidic treatment. On the contrary, the transformation by water gain occurs in vermiculites treated with hydrogen peroxide and in those subjected to ionic metal exchange.

Structural changes of vermiculites induced by the abovementioned treatments provide evaluable information on the relationship between the structure of vermiculite and their industrial applications. The said relationship would allow predicting the structural order-disorder of the vermiculite, the obtaining of purest vermiculite, or the environmental fate of toxic metals.

The changes suffered by vermiculites due to the treatments applied could give light to ambiguities about their geological origin and hydrothermal and/or supergene processes. Early field and laboratory evidence and current experiments showing changes in vermiculites caused by treatment with hydrogen peroxide and ion-metal exchange, with water gain, could point to a supergenic origin. Regarding the treatments that involve water loss in vermiculites, it is not discarded that a more detailed study helps to reveal data related to the hydrothermal origin.

Conflict of interest

The author declares that she has no known competing financial interests or personal relationships that could have appeared to influence the work reported in this chapter.

Author details

Celia Marcos
Geology Department, Oviedo University, Oviedo, Spain

*Address all correspondence to: cmarcos@uniovi.es

References

[1] Mathieson AM, Walker GF. Crystal structure of magnesium-vermiculite. American Mineralogist. 1954;**39**:231-255

[2] Calle C de la, Suquet H and Pons CH: Stacking order in a 14.30 Å Mg-vermiculite, Clays and Clay Minerals, 1988;**36**:481-490

[3] Walker GF. Mechanism of dehydration of Mg-vermiculite. Clays and Clay Minerals. 1956;**4**:101

[4] Vali H, Hesse R. Identification of vermiculite by transmission electron microscopy and x-ray diffraction. Clay Minerals. 1992;**27**:185-192

[5] Collins DR, Fitch AN, Catlow RA. Dehydration of vermiculites and montmorillonites: a time-resolved powder neutron diffraction study. Journal of Materials Chemistry. 1992;**8**:865-873

[6] Reichenbach HG, Beyer J. Dehydration and rehydration of vermiculites: IV. Arrangements of interlayer components in the 1.43 nm and 1.38 nm hydrates of Mg-vermiculite. Clay Minerals. 1994;**29**:327-340

[7] Reichenbach HG, Beyer J. Dehydration and rehydration of vermiculites: II. Phlogopitic Ca-vermiculite. Clay Minerals;**30**:273-286

[8] Reichenbach HG, Beyer J. Dehydration and rehydration of vermiculites: III. Phlogopitic Sr. and Ba-vermiculite. Clay Minerals. 1997, 1995;**32**:573-586

[9] Marcos C, Argüelles A, Ruíz-Conde A, Sánchez-Soto PJ, Blanco JA. Study of the dehydration process of vermiculites by applying a vacuum pressure: Formation of interstratified phases. Mineralogical Magazine. 2003;**67**(6):1253-1268. DOI: 10.1180/0026461036760163

[10] Marcos C, Arango YC, Rodríguez I. X-ray diffraction studies of the thermal behaviour of commercial vermiculites. Applied Clay Science. 2009;**42**:368-378. DOI: 10.1016/j.clay.2008.03.004

[11] Suzuki M, Wada N, Hines DR, Whittingham MS. Hydration states and phase transitions in vermiculite intercalation compounds. Physical Review B. 1987;**36**(5):2844-2851

[12] Ruiz-Conde A, Ruiz-Amil A, Pérez-Rodríguez JL, Sánchez-Soto PJ. Dehydration-rehydration in magnesium vermiculite: Conversion from two-one and one-two water hydration states through the formation of interstratified phases. Journal of Materials Chemistry. 1996;**6**:1557-1566

[13] Midgley HG, Midgley CM. The mineralogy of some commercial vermiculites. Clay Minerals Bulletin. 1960;**23**:142-150

[14] Couderc P, Douillet P. Les vermiculites industrielles: Exfoliation, caractéristiques mineralogiques et chimiques. Bulletin de la Société Francaise de Céramique. 1973;**99**:51-59

[15] Hillier S, Marwa EMM, Rice CM. On the mechanism of exfoliation of 'Vermiculite'. Clay Minerals. 2013;**48**:563-582. DOI: 10.1180/claymin.2013.048.4.01

[16] Myers JB. Vermiculite. In: Industrial Minerals and Rocks. New York: American Institute of Mechanical Engineers; 1963. pp. 889-895

[17] Mamina AK, Koteĺnikova EN, Muromtsev VA. Influence of the structural perfection of phlogopite crystals on their cleavability by hydrogen peroxide. Inorganic Materials. 1990;**26**:2104-2107

[18] Suquet H, Chevalier S, Marcilly C, Barthomeuf D. Preparation of porous

materials by chemical activation of the Llano vermiculite. Clay Minerals. 1991;**26**:49-60

[19] Üçgül E. Thermal and chemical exfoliation characteristics of Sivas-Yldzeli-Karakoç phlogopite [MSc thesis]. Turkey: Hacettepe University of Ankara; 1997 (in Turkish)

[20] Üçgül E, Girgin IÇ. Chemical exfoliation characteristics of Karakoç phlogopite in hydrogen peroxide solution. Turkish Journal of Chemistry. 2002;**26**:431-439

[21] Obut A, Girgin IÇ. Hydrogen peroxide exfoliation of vermiculite and phlogopite. Minerals Engineering. 2002;**15**:683-687. DOI: 10.1016/S0892-6875(02)00161-9

[22] Weiss Z, Valášková M, Seidlerová J, Šupová-Křístková M, Šustai O, Matějka V, et al. Preparation of vermiculite nanoparticles using thermal hydrogen peroxide treatment. Journal of Nanoscience and Nanotechnology. 2006;**6**(3):726-730. DOI: 10.1166/jnn.2006.116

[23] Marcos C, Rodríguez I. Expansion behaviour of commercial vermiculites at 1000°C. Applied Clay Science. 2010;**48**:492-498. DOI: 10.1016/j.clay.2010.02.012

[24] Marcos C, Rodríguez I. Expansibility of vermiculites irradiated with microwaves. Applied Clay Science. 2011;**51**:33-37. DOI: 10.1016/j.clay.2010.10.019

[25] Justo A, Pérez-Rodríguez JL, Sánchez-Soto PJ. Thermal studies of vermiculites and mica-vermiculite interstratifications. Journal of Thermal Analysis. 1993;**40**:59-65

[26] Shiqing G, Kang X, Wang L, Lichtfouse E, Wang C. Clay mineral adsorbents for heavy metal removal from wastewater: A review. Environmental Chemistry Letters. 2019;**17**(2):629-654. DOI: 10.1007/s10311-018-0813-9

[27] Valášková M, Martynková GS. Vermiculite: Structural properties and examples of the use. In: Valášková M, Martynková GS, editors. Clay Minerals in Nature - Their Characterization, Modification and Application. IntechOpen; 2012. DOI: 10.5772/51237

[28] Suzuki M, Suzuki IS. Superparamagnetic behavior in a Ni vermiculite intercalation compound. Physical Review B. 2001;**64**:104418. DOI: 10.1103/PhysRevB.64.104.418

[29] Hindman JR. Vermiculite. In: Kogel JE, Trivedi NC, Krukowsky ST, editors. Industrial Minerals and Rocks: Commodities, Markets, and Uses. Colorado: Society for Mining, Metallurgy, and Exploration; 2006. pp. 1015-1027

[30] Bergaya F, BKG T, Lagaly G, editors. Developments in Clay Science. In: Handbook of Clay Science. 2nd ed. Vol. 5. Elsevier Ltd.; 2006

[31] Klein C, Dutrow B. Manual of Mineral Science. 23rd ed. New York: Wiley; 2007. p. 716

[32] Abollino O, Giacomino A, Malandrino M, Mentasti E. Interaction of metal ions with montmorillonite and vermiculite. Applied Clay Science. 2008;**38**(3-4):227-236. DOI: 10.1016/j.clay.2007.04.002

[33] Zhang K, Xu J, Wang KY, Cheng L, Wang J, Liu B. Preparation and characterization of chitosan nanocomposites with vermiculite of different modification. Polymer Degradation and Stability. 2009;**94**(12):2121-2127. DOI: 10.1016/j.polymdegradstab.2009.10.002

[34] Marcos C, Argüelles A, Khainakov SA, Rodríguez-Fernández J, Blanco JA.

Spin-glass freezing in Ni-vermiculite intercalation compound. Journal of Physics. Condensed Matter. 2012;**24**(34):346001-346010. DOI: 10.1088/0953-8984/24/34/346001

[35] Marcos Pascual C, Argüelles A, Leoni M, Blanco JA. Location of Ni^{2+} in nickel-intercalated vermiculites. Applied Clay Science. 2014;**91-92**:79-86. DOI: 10.1016/j.clay.2014.01.017

[36] Brigatti MF, Colonna S, Malferrari D, Medici L, Poppi L. Mercury adsorption by montmorillonite and vermiculite: A combined XRD, TG-MS, and EXAFS study. Applied Clay Science. 2005;**28**:1-8. DOI: 10.1016/j.clay.2004.03.006

[37] Jiménez de Haro MC, Pérez-Rodríguez JL, Poyato TJ, Pérez-Maqueda LA, Ramirez-Valle V, Justo A. Effect of ultrasound on preparation of porous materials from vermiculite. Applied Clay Science. 2005;**30**:11-20. DOI: 10.1016/j.clay.2005.02.004

[38] Hennies WT, Stellin A Jr. A jazida de vermiculita de Piaui, Estado do Piauí. Anais de Congressos Brasileiros de Geologia. 1978;**4**:1796-1804

[39] González García F, García Ramos G. On the genesis and transformations of vermiculite. In: Transactions 7th International Congress of Soil Science. Vol. 4. Madison, Wisconsin; 1960. pp. 482-491

[40] Velasco F, Casquet C, Ortega Huerta M, Rodríguez Gordillo J. Indicio de vermiculita en el skarn magnético (aposkarn flogopítico) de La Garrenchosa. Vol. 2. Sta. Olalla, Huelva, Sociedad Española de Mineralogía; 1981. pp. 135-149

[41] Justo A, Estudio físico-químico y mineralógico de vermiculitas de Andalucia y Badajoz [thesis]. Spain: Sevilla University; 1984

[42] Luque FJ, Rodas M, Doval M. Mineralogía y Génesis de los yacimientos de vermiculita de Ojen. Boletín de la Sociedad Española de Mineralogía. 1985;**8**:229-238

[43] López González JD, Barrales Rienda JM. Caracterización y propiedades de una vermiculita de Benahavis (Málaga). Anales de Química. 1972;**68**:247-262

[44] Justo A. Estudio físico-químico y mineralógico de vermiculitas de Andalucía y Badajoz [thesis]. Spain: Sevilla University; 1984

[45] Justo Erbez A, Pérez Rodríguez JL, Maqueda C. Estudio mineralógico de una vermiculita de Ojén (Málaga). Boletín de la Sociedad Española de Mineralogía. 1983;**16**:59-67

[46] Marcos C, Rodriguez I. Vermiculites irradiated with ultraviolet radiation. Applied Clay Science. 2015;**109**:127-135. DOI: 10.1016/j.clay.2015.03.002

[47] Argüelles A, Khainakov SA, Rodríguez-Fernández J, Leoni M, Blanco JA, Marcos C. Chemical and physical characterization of iron-intercalated vermiculite compounds. Physics and Chemistry of Minerals. 2011;**38**:569-580. DOI: 10.1007/s00269-011-0429-0

[48] Leoni M, Gualtieri AF, Roveri N. Simultaneous refinement of structure and microstructure of layered materials. Journal of Applied Crystallography. 2004;**37**:166-173. DOI: 10.1107/S0021889803022787

[49] Marcos C, Adawy A, Rodríguez I. Unpublished

[50] Ravichandran J, Sivasankar B. Properties and catalytic activity of acid-modified montmorillonite and vermiculite. Clays and Clay Minerals. 1997;**45**:854-858

[51] Wypych F, Adad LB, Mattoso N, Marangon AAS, Schreiner WH. Synthesis and characterization of disordered layered silica obtained by selective leaching of octahedral sheets from chrysotile and phlogopite structures. Journal of Colloid and Interface Science. 2005;**283**:107-112. DOI: 10.1016/j.jcis.2004.08.139

[52] Komadel P, Madejova J. Acid activation of clay minerals. In: Bergaya F, Lagaly G, editors. Handbook of Clay Science (Developments in Clay Science). Vol. 5. 2013. pp. 385-409. DOI: 10.1016/B978-0-08-098258-800.013-4

[53] Chmielarz L, Wojciechowska M, Rutkowska M, Adamski A, Węgrzyn A, Kowalczyk A, et al. Acid activated vermiculites as catalysts of the DeNOx process. Catalysis Today. 2012;**191**:25-31. DOI: 10.1016/j.cattod.2012.03.042

[54] Marcos C, Rodriguez I. Some effects of trivalent chromium exchange of thermo-exfoliated commercial vermiculite. Applied Clay Science. 2014;**90**:96-100. DOI: 10.1016/j.clay.2014.12.032

[55] Marcos C, Medoro V and Adawy A. Unpublished

[56] Marcos C, Rodriguez I. Exfoliation of vermiculites with chemical treatment using hydrogen peroxide and thermal treatment using microwaves. Applied Clay Science. 2014;**87**:219-227. DOI: 10.1016/j.clay.2013.11.008

[57] Marcos C, Rodriguez I. Structural changes on vermiculite treated with methanol and ethanol and subsequent microwave irradiation. Applied Clay Science. 2016;**123**:304-314. DOI: 10.1016/j.clay.2016.01.024

[58] Marcos C, Rodriguez I. Effect of propanol and butanol and subsequent microwave irradiation on the structure of commercial vermiculites. Applied Clay Science. 2017;**114**:104-114. DOI: 10.1016/j.clay.2017.05.014

[59] Marcos C, Adawy A, Río del Z; Unpublished

[60] Stawiński W, Freitas O, Chmielarz L, Węgrzyn A, Komędera K, Mordarski G, et al. The influence of acid treatments over vermiculite based material as adsorbent for cationic textile dyestuffs. Chemosphere. 2016;**153**:115-129. DOI: 10.1016/j.chemosphere.2016.03.004

[61] Mercier L, Detellier C. Preparation, characterization, and applications as heavy metals sorbents of covalently grafted thiol functionalities on the interlamellar surface of montmorillonite. Environmental Science and Technology. 1995;**29**:1318

[62] Kraepiel AML, Keller K, Morel FMM. A model for metal adsorption on montmorillonite. Journal of Colloid and Interface Science. 1999;**210**:43-54

[63] Malandrino M, Abollino O, Giacomino A, Aceto M, Mentasti E. Adsorption of heavy metals on vermiculite: Influence of pH and organic ligands. Journal of Colloid and Interface Science. 2006;**299**:537-546. DOI: 10.1016/j.jcis.2006.03.011

[64] da Fonseca MG, de Oliveira MM, Arakaki LNH, Espinola JGP, Airoldi C. Natural vermiculite as an exchanger support for heavy cations in aqueous solution. Journal of Colloid and Interface Science. 2005;**285**:50-55. DOI: 10.1016/j.jcis.2004.11.031

[65] Derkowski A, Drits VA, McCarty DK. Rehydration of dehydrated-dehydroxilated smectite in a low water vapor environment. American Mineralogist. 2012;**97**:110-127. DOI: 10.2138/am.2012.3872

[66] Huo X, Wu L, Liao L, Xia Z, Wang L. The effect of interlayer cations on the expansion of vermiculite. Powder Technology. 2012;**224**:241-246. DOI: 10.1016/j.powtec.2012.02.059

[67] Folorunso O, Dodds C, Dimitrakis G, Kingman S. Continuous energy efficient exfoliation of vermiculite through microwave heating. International Journal of Mineral Processing. 2012;**114-117**:69-79. DOI: 10.1016/j.minpro.2012.10.003

[68] Xue Y, Zhang S, Yang W. Influence of expanded vermiculite on fire protection of intumescent fireproof coatings for steel structures. Journal of Coatings Technology and Research. 2015;**12**(2). DOI: 10.1007/s11998-014-9626-3

[69] Zacarias RA, Forte GS, Fontgalland G, Carvalho JN, Idalmir SQ. Thermal analysis of vermiculite using microwave. In: International Instrumentation and Measurement Technology Conference (I2MTC), 14-17 May 2015. Houston, TX, USA: IEEE; 2018. pp. 1-5

[70] Kaur S, Singh S, Singh L. Physio-chemical properties of gamma-irradiated vermiculite and their significance for radiation protection and thermoluminescence. American Mineralogist. 2014;**99**:2018-2024. DOI: 10.2138/am-2014-4873

[71] Kawamura H, Taniguchi T. Spin Glasses. In: Buschow KHJ, editor. Handbook of Magnetic Materials. Vol. 24. Elsevier; 2015. pp. 1-137. DOI: 10.1016/bs.hmm.2015.08.001

[72] Santos SSG, Silva HRM, de Souza AG, Alves APM, da Silva Filho EC, Fonseca MG. Acid-leached mixed vermiculites obtained by treatment with nitric acid. Applied Clay Science. 2015;**104**:286-294. DOI: 10.1016/j.clay.2014.12.008

[73] Stawiński W, Węgrzyn A, Stawiński W, Freitas O, Chmielarz L, Mordarski G, et al. Simultaneous removal of dyes and metal cations using an acid, acid–base and base modified vermiculite as a sustainable and recyclable adsorbent. The Science of the Total Environment. 2017;**576**:398-408. DOI: 10.1016/j.scitotenv.2016.10.120

[74] Węgrzyn A, Stawiński W, Freitas O, Komędera K, Błachowski A, Jęczmionek Ł, et al. Study of adsorptive materials obtained by wet fine milling and acid activation of vermiculite. Applied Clay Science. 2018;**155**:37-49. DOI: 10.1016/j.clay.2018.01.002

[75] Motokawa R, Endo H, Yokoyama S, Nishitsuji S, Kobayashi T, Suzuki S, et al. Collective structural changes in vermiculite clay suspensions induced by cesium ions. Scientific Reports. 2014;**4**:6585. DOI: 10.1038/srep06585

[76] Bassett WA. The geology of vermiculite occurrences. Clays and Clay Minerals. 1961;**10**(1):61-69

[77] Toksoy-Koksal F, Turkmenoglu AG, Goncuoglu MC. Vermiculitization of Phlogopite in metagabbro, Central Turkey. Clays and Clay Minerals. 2001;**49**(1):81-91. DOI: 10.1346/CCMN.2001.0490107

[78] Churchman GJ, Lowe DJ. Alteration, formation, and occurrence of minerals in soils. In: Huang PM, Li Y, Summer ME, editors. Handbook of Soil Sciences, 2nd Properties and Processes. Boca Raton, FL: CRC Press (Taylor & Francis); 2012. pp. 20.71-20.72

[79] Mohammed IJ, Al-Mashaikie SZ. Origin and distribution of clay minerals in the mudstones of the kolosh formation in rawandoz area, northeastern Iraq. Iraqi Geological Journal. 2018;**51**(2):75-90

Chapter 4

Adsorption of Heavy Metals by Microwave Activated Shale/ Asphaltite Char/Zeolite Granule Composts from Hazardous Sludges and Industrial Waste Slurries

Yıldırım İsmail Tosun

Abstract

There is a great concern about surface water pollution with high level mercury, lead (Pb) over 10 mg/l, 30 mg/l to the fishing lakes and streams in Şırnak Province even contaminating fresh water fishing and poisonening of human by merury and lead in thr region. The chromium over 50 mg/l from industrial seepages was disposed to lakes and streams in our country. There is a great green concern prompting land in order to control acidic mine waters so that the research study controlled and avoided hazardous metal limits of residual stream contaminants of heavy metals by sorption local clay and zeolite compost. The contamination rate changes to those based on seepage concentrations and wetness. The stream amendments, such as shale char carbonized from Şırnak asphaltite containing 52–60% shale activated by acid washing under microwave radiation as geo material composted for waste water treatment should control contaminated effluents concentration. The field studies to evaluate the stability of heavy metal concentrations and salts were scarce. The initial objective of this study was to determine the effects of seepage flow to surface and groundwater from the industrial discharge. In this study, important investigations have been made on composite granules production with Şırnak shale char and zeolite feed in order to activated in microwave oven 2 M HCl dissolution. The compost sorbent for high level heavy metal sorption in laboratory water packed bed column adsorption compost system. However, the results of filled packed bed zeolite yield high metal transfer to compost. Due to the complex chemistry of shale pores, and high porosity, heat conduction improved in the microwave sorption depended on granule size decreased. The other heavy metal sorption distribution was changed in the activation dependent on the microwave heating power.

Keywords: zeolite, microwave radiation, salt slurries, metal sorption, energy toxic risk assessment, stochastic cost estimation, treatmen sorbent simulation, hybrid sorbent, waste sludge, salt slurries, microwave activation waste water treatment, heavy metal, zeolite composts, shale

1. Introduction

This investigation of water treatment firstly proposed to control mud in acidic manner even dissolving hazardous coal mine cadmium, lead and mercury can be extremely hazardous in fresh water source contamination even ındustrial waste water management. The contamination research by soil remediation was existing in water logged areas. The strategical problems of water contamination and treatment by different type clay composts and the quality of them mostly exist in the irrigated areas like in South Eastern rocky plains of Şırnak and Batman, Turkey. The climate change and ground water changes generally resulted in over irrigation, seepage losses through channel and distributions, contaminated water control, management practices and inadequate control of drainage system. Analysis of high water table in water logged areas and drainage of irrigated areas have not been paid adequate attention in the planning and management of water resources, partly due to lack of requisite data and partly due to flood and rainfall in the country. In order to develop suitable water management strategies and controlling the extent of water logging in the area. GIS may facilitate the reconstruction of the ecological environment but also to accommodate the sustainable development of the water resources and waste water. In this study, the control of lakes and streams by hydrological characteristics of the Batman city were explained and the effect of soil characteristics on the the city was examined. In the investigation, hydrological features and the urbanization with new settlements need modeling regarding available water source. The hydrological property of settlement areas with dense populated areas in the model was determined by Geographic Information Systems (GIS) techniques. The main purpose of this study is to investigate the effect of settlement on the basic hydrological structure by studying the characteristics of the ground topography, ground water elevation, slope and viewing. GIS techniques were used in the creation of the thematic maps and in the analysis of the parameters. Finally, the GIS study models created, the available water source change and a stream contamination model was provided sufficient source control at the Batman province. The presence of this stream and lake contamination, soil structure in the Batman province reveals the contamination by acidic mine waters and potential flood scale and flood risk. This study produced more hazardous contamination data with hydrological streams dicharged to lakes and streams with GIS. GIS has made it possible to obtain more qualified data by enabling the use of waste water treatment by mobil units in this research (**Figure 1a** and **b**) [1, 2].

The hydrological studies carried out by Water Association showed that high level Tigris river stream flows from 650 m level attitude through 610 m−600 m levels laguuns in Batman province with hig risk of flood (**Figure 2**).

The chalky limestone layers of Batman province reveals high amount of water suddenly at the stream below levels to 2 m high over agricultural wheat land and the potential flood risk line [2] (**Figure 3**) [1–3].

1.1 Accumulation of contaminating acidic waters, heavy metals and distribution in floods and laguuns near dams

The oxygen content and and electroopotantial of waters adequately accounted in stream flows causing animal feed contamination n the pastoral fields by soil and growing grass near by this contaminated stream laguuns.

In the hot streams and acidic mine waters the ferric iron and sulfate tend to be highly common as AMD seepage, alkali resulting from the reduction of these two species, a weak base (bicarbonate) and producing astrong base (hydroxyl ions), also generate net alkalinity (Eq. (8)). The indirect acid production was relatively high

DOI: http://dx.doi.org/10.5772/intechopen.94404

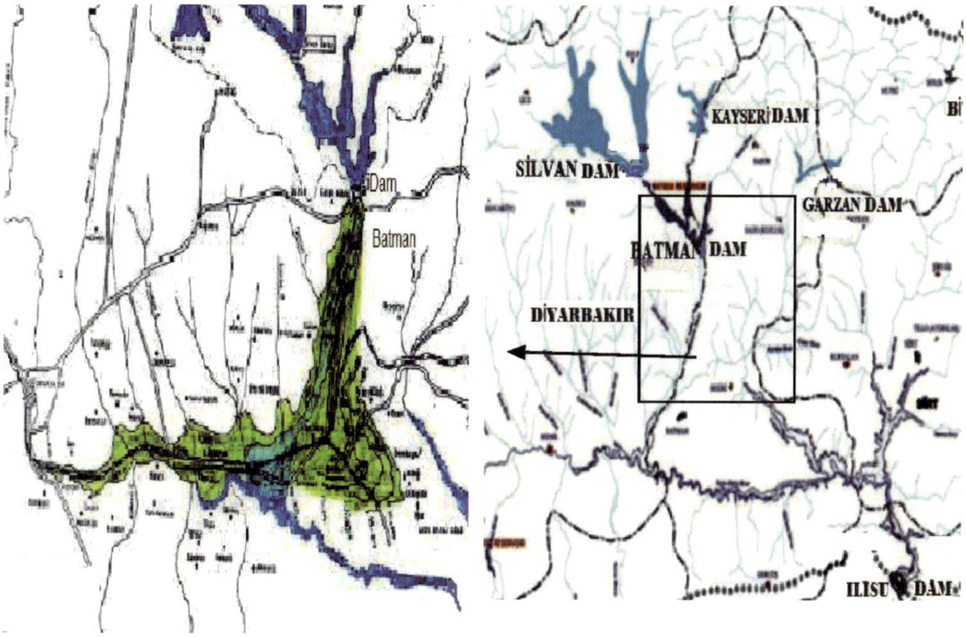

Figure 1.
(a) Groundwater Flows and Hydrological Stream Discharges in Batman Province at scale of 1/20000 (b) 1/50000.

Figure 2.
(a) Satellite view of Batman City and province reveals the potential flood scale 1/20000 and (b) earth view with flood risk water lines of Batman City and province 1/20000.

rate at higher pH levels over 5 with dissolution of heavy metals in sulphide minerals and neutralization by alkali matters govern the dissolution by the reactions as given below [4–6];

Bicarbonate sulphate hot streams reduce aciditiy as below Eq. (1).

$$SO_4{}^{=} + 2HCO_3{}^{=} + H_2O + 2H^+ \rightarrow H_2S + 3CO_2 + 2H_2O \quad (1)$$

The sulfide produced is strongly reactive towards heavy metals as given Eqs. (2) and (3):

$$Fe^{2+} + H_2S \rightarrow FeS + 2H^+, \quad (2)$$

$$Zn^{2+} + H_2S \rightarrow ZnS + 2H^+, \quad (3)$$

Figure 3.
Batman province some set reveals the potential flood ban and flood risk.

which form very insoluble sulfide compounds. FeS is unstable relative to pyrite and the further reaction, which is an oxidation of S^{2-} to S^{-}, as given in Eq. (4);

$$FeS + S \rightarrow FeS_2 \quad (4)$$

generated in the late muds close to the settled mud - water interface. ZnS and PbS, in the sulphide complex structure are much stable and retain S in the -2 state. However, the sulfate part of reaction Eq. (5) [7, 8] may cause redox effect an oxidation. Then.

$$H_2S + 4H_2O \rightarrow {SO_4}^{2-} + 10H^{+} + 8e^{-}. \quad (5)$$

The aerated and oxygen rich waters oxidizing sulphidic character metal precipitates to sulphate and chloride dissolution by unstable forms, but over ph 9 as shown in Figure electropotantial matter of waste waters provides hydroxide precipitates in soil mud. Even jarosite form precipitates occuuring in hot water streams area with redish brownish precipitates, however those type resıduals stuck over sand may become sweet salty alg fish feed even causing higher heavy metal contamination for fish farming and stream fishing. Batman province copper and lead sulphide deposits and hot streams of high sulphate come out high potential contamination [7–13] of fresh waters soueces at pH Eh diagram stability as given in **Figure 4a** and **b** folowing flood.

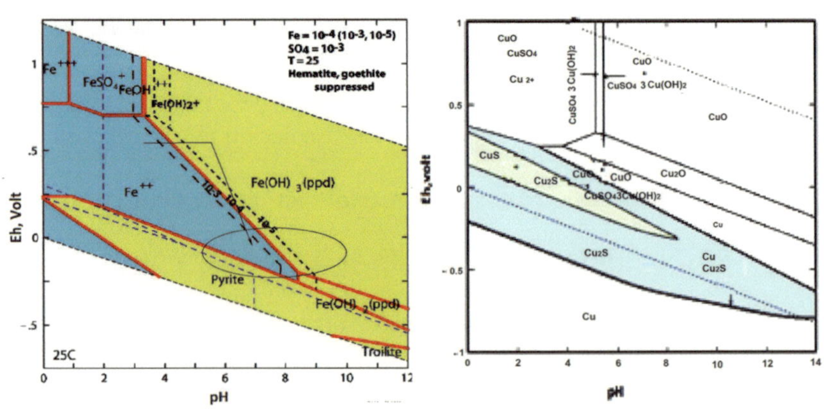

Figure 4.
(a) Eh-pH diagrams for metal Fe and (b) Eh-pH diagrams for metal Cu stability.

The dissolution kinetics of soil mud particle for Pb heavy metal is followed by equation

$$\frac{dc_{Pb}}{dt} = k_i e^{-tic} dc \tag{6}$$

Where cPb Lead contamination mg/l, k the rate of dissolution of lead, i is the reaction style, t is time,

The dissolution concentration of accumulated metal in aliquate of lake streams as regarding Pb heavy metal contamination is followed by equation, where n is kinetic order type

$$\frac{dc_{Pb}}{dt} = k_i c^{tin} dc \tag{7}$$

The dissolution concentration of accumulated metal in aliquate of sulfurous hot water streams as regarding Pb heavy metal contamination is followed by equation, where SO_4^{-2} sulphate concentration in effluent. f_i is concentration rate of sulphate

$$\frac{dc_{Pb}}{dt} = k_i c^{tin}.dc.f_i\left(SO_4^{-2}\right)^{tin} \tag{8}$$

The dissolution concentration of accumulated metal in aliquate of limestone rocks dissolution by hot water streams in subground lakes with high CO2 gas dissolved streams as regarding Pb heavy metal contamination is followed by equation, where HCO_3^{-2} bicarbonate concentration in effluent

$$\frac{dc_{Pb}}{dt} = k_i c^{tin}.dc.f_i\left(HCO_3^{-2}\right)^{tin} \tag{9}$$

The dissolution concentration of accumulated metal in aliquate of high fertilizer dissolution by wrong amount of fertilizer use in theagricultural fields discharged to streams as regarding Pb heavy metal contamination is followed by equation, where HNO_3^{-2} nitrate concentration in effluent

$$\frac{dc_{Pb}}{dt} = k_i c^{tin}.dc.f_i\left(HNO_3^{-2}\right)^{tin} \tag{10}$$

The dissolution rates of heavy metals in acidic mine waters and sulfurous hot streams occurred in the region of Ilısu Dam, Güçlükonak, Şırnak and Batman Province sites. Fish farming require below 1 mg/l Pb Cu and Cd and Zn in which basaltic rocks and copper ore deposites contained highly around %1-2Pb and 200 mgCu at high attitude deposits in Siirt and Şırnak. The contamination of some accumulated heavy metal contents of hot streams and soils in the region are given in **Table 1**.

Urbanization and economic growth in the high populated Batman city evolved along with the management of natural resources. In this process, provision of drinking water supply and distribution service for urban areas also developed on the same plane. The effective role of the public was felt in meeting the water resources management and service. Infrastructure investments are centrally located, water resources are found, structured, stored, distributed and refined. Large investments have been made in order to meet the fresh water need. The use of water resources (water withdrawal and level shift) and evaluation for development and community needs have been studied. However, the amount and quality of water that the

Effluent, mg/l	Şırnak coal mine pool	Şırnak, hezil stream	Güçlükonak hot stream	Batman hot stream	Şırnak kasrik laguun	Ilısu dam laguun1	Ilısu Dam Laguun2
Hg	8,11	4,71	12,3	14,11	4,71	4,71	4,71
Pb	10,58	14,53	23,2	12,58	11,53	5,7	5,2
Fe	40,33	70,62	59	93,3	56,2	60,62	67,62
K + Na	7,52	8,46	8,7	8,52	8,6	≥70	≥50
Cd	24,72	19,56	14,1	14,72	19,56	16	15
Mn	33,3	24,1	24,2	43,3	24,1	≤25	≤25
Cu	27,2	30,2	15,7	7,2	10,2	≤15	≤15
As	1,10	2,44	2,8	2,10	2,44	≤5	≤5
SO_4	0,57	0,37	1,9	0,67	0,55	≤15	≤15
Soil,ppm							
Hg	34,11	48,71	52,3	54,11	40,71		
Pb	10,58	24,53	23,2	20,58	11,53		
Fe	4,33	7,62	5,9	9,33	5,62		
K + Na	74,52	81,46	81,7	84,52	88,6	≥70	≥50
Cd	24,72	9,56	10,1	4,72	19,56		
Mn	2,72	3,02	1,5	0,72	1,02	≤5	≤5
Cu	3,33	2,41	2,4	4,33	2,41	≤5	≤5
As	1,10	2,44	2,8	2,10	2,44		
SO_4	0,57	0,37	1,9	0,67	0,55		

Table 1.
Şırnak and Batman province reveals the potential contamination scale and high contamination risk of fresh water source with flood.

ecology need is not addressed. Everything is built on the theme of "develop-supply-use". Parameters considered in the planning of water resources were population estimate, per capita water demand, fish farming, agricultural production, economic productivity level.

1.2 Fish farming in water lakes and streams

The hazardous high level contaminants occurred in the such acidic seepages or acidified chelate mixing to streams should be neutralized by oxidizing reagents such as ozone or neutralizing alkaline washing so that resulted effluent contamination by Hg, Pb, Cr, Cd, Cu, Zn, Fe, SO_4 rates were so low. The oxidation recycling of residual contaminants was not a serious threat using these parameters, future water demand forecasts are used and these estimated values are used when designing the systems to meet the demand. In this approach, the demand for water has been determined independently of the specific needs of fish farming, agricultural irrigation, and human needs, the amount of water a healthy ecosystem will need, or actual regional water availability. The next step in traditional planning is to identify projects that will reduce the gap between estimated water supply and demand. In every scale, the planning action (region, basin, city) is used for the regular and healthy spatial development uses (housing, commerce, industry, recreation, other social) in the metropolitan cities which are especially migrating in our country and in medium size settlements Such as equipment as directed by location decisions; It also determines the water demand of the city at the same time with its population and density of buildings and its quality and quantity of usage. While city plans shape the socio-economic and physical structure of the city, with the proposed land use, employment, population and density decisions, the city's daily water demand is also shaped. Therefore, any kind of urban development outside the plan creates an unhealthy environment that affects the quality of life of the city, as well as poses a serious threat to the water resources (increased water consumption pressure and pollution) (Urban Planning Chamber Water Commission, 2006). Survey, planning (feasibility) and Project work will be given efficiency. The quality of the water quality will be preserved, improved and monitored. Heavy metal contamination hazard maps will be prepared and an early warning system will be established.

Nowadays, oil, grease and other pollutants as absorbent material in cleaned, bleach mud in food industry, in pharmacy adsorbant, as a catalyst carrier in the chemical industry and in many other industries it is used for various purposes. Clay bentonite, which is used as industrial absorbant, sepiolite, atapulgite and kaoline. This clay is good absorbant or active bentonite or montmorillonite. it is highly demand for absorbant substrate for fresh fruit drinks and brewery, water tretmennts in Europe. The consumption has increased to around 5 million tonnes/year in the 2020s while in 1994 it exceeded 2 million tons/year. In terms of bentonite and sepiolite, it is known that it has large beds and it is enough on these beds until recently research work was arried out for heavy metalcontamination. In this study, bentonite and other clays, shale and marly shale of Şırnak and absorbance properties, areas of use, production and market conditions. The bulk density of absorbant clay bentonite and atapulgite, the amount of moisture and the absorbant capacity. Bentonite and atapulgite absorbance by passing through certain processes was performed and the absorbance was measured at the mechanical strength change has been studied.

In general, a grainy and fine grained the raw materials described as the shale, kaoline, illite, marly chlorite and smectite alteration coverings of rich rock abrasion and carry over fresh water occurs in the basins. Kaolinite, montmorillonite, illite, chlorite, sepiolite and atapulgiteone or several quartz, cristobalite, amphibole, feldspar, calcite, magnesite, dolomite, gypsum, alunite, and natural clay containing one or more minerals heterogeneous mixtures [7]. Mineral depending on their content and chemical composition As the color of the killer, white, pink, gray, green, in various shades of yellow, blue and brown [8]. The chemical analyzes indicate that the killer is mainly silica, alumina and water can be distinguished in most cases iron, alkali and alkaline earth in quantities. In this study, montmorillonite group bentonite and chain clay atapulgite used.

DTA curve and the water away from the cation between the layers. In addition, endothermic DTA peaks originating from the removal of H-bound shale in the samples were observed at approximately 200°C, and endothermic peaks of hydrate bound to the Brønsted centers were observed at approximately 300°C.

The fact that clay and clay minerals, which constitute a significant part of the ground and underground resources of our country, are not processed sufficiently is an important issue that causes serious economic losses for our country. In order to produce clay minerals in high quality and desired properties, some processes such as acid activation, organo-clay preparation, microwave dissolution, calcination and cation exchange are used [12–15]. Acid-activated clay, known as bleaching earth, is used in scientific research as a selective retainer, catalyst, catalyst support and in the differentiation of the killer [16]. Bentonite with desired surface properties, porosity and hence retention capacity is mainly produced by dry or wet acid activation using mineral acids such as H_2SO_4 and HCl [17]. The main purpose in acid activation is to reach the desired structure without disrupting the layered crystal structure of the clay. For this reason, the acid/clay ratio, temperature, acidity, acid concentration, type and duration of activation, kiln type and physical properties and amount of washing water are important considerations to be taken into account when performing the appropriate activation.

1.3 Zeolite - clay compost with brønsted and lewis acid centers

The Brønsted acid centers are mainly associated with the inner layer zone and the Lewis acid centers with clay marginal surfaces. The water molecules in the spheres surrounding the exchangeable cations are protonic depending on the degree of polarization of the metal cation and behave as Brønsted acid. In addition, the surface silanol groups (Si-OH) resulting from the breakage of the Si-O-Si bonds in the tetrahedral layer in the killer contribute to the Brønsted acid centers. Lewis acid centers are also associated with the co-shaped exchange of the Al^{3+} and Mg^{2+} cations in the octahedral layer and the Si^{4+} and Al^{3+} and Fe^{3+} cations in the tetrahedral layer, albit associated (**Figure 5**) [18] with metal atoms on the crystal edges. The oxygen planes in the space between the plates act as a pair of electrons, ie Lewis bases. The Hammett acid indicator technique, the n-butyl amine back titration technique and the investigation of the attachment geometry of the species such as cyclohexylamine, n-butylamine and pyridine can be used to determine the species and amount of acid centers in clay minerals rapidly [18, 19].

The increasing demand of bentonite utilization for advanced material technology and the limited reserves of high quality bentonites push the reserachers and the operators/producers to evaluate the lower quality calcium and mixed bentonites for the replacement of Na-bentonites in use. The technological properties of bentonites, however, can be upgraded by the application of concentrating and alkali activation. Mostly, wet concentration methods such as decantation, hyrocycloning and centrifuging have been applying and water quality and ion type/amount which the water carries becomes more important to controll the further activation process

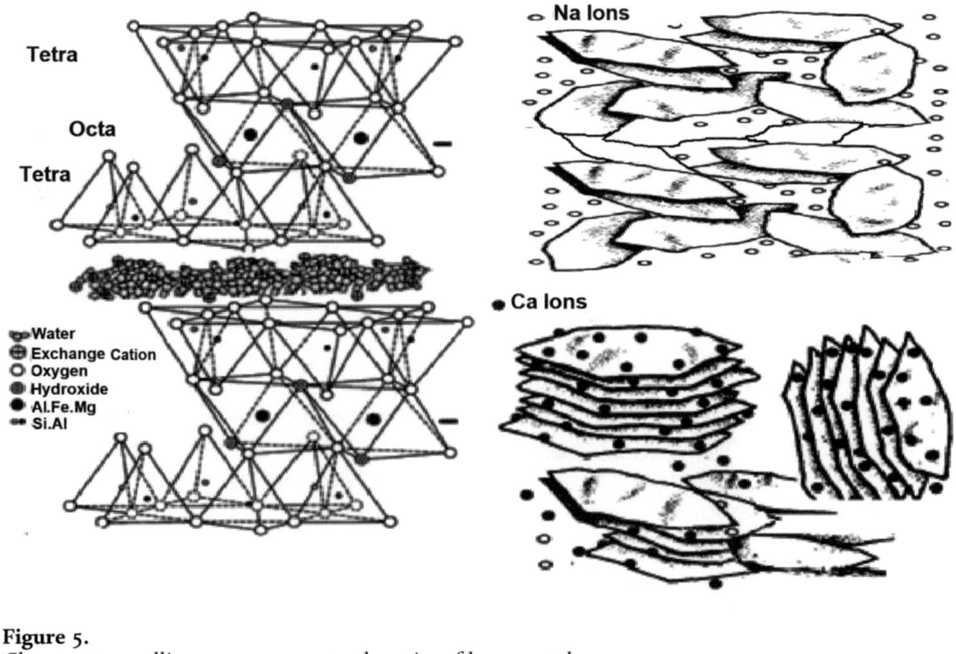

Figure 5.
Clay structure welling manner prompts adsorption of heavy metals.

since bentonites carry the releasable and exchangable cations on interlayers which interact with ions in water.

In this study, the effect of water quality (ion type and amount in water) was subjected to the concentration and further alkali activation tests with mixed type bentonite received from Resadiye/Tokat bentonite deposit. Deionized, filtered and tap water and syntetic water including different salts namely $CaCl_2.2H_2O$, NaCl, $MgCl_2$, KCl, $FeCl_3$. $6H_2O$ were used as separation media in concentration by settling and decantation. The effect of water quality on concentration and alkali activation were declared based on the pH, CEC (Cation exchange capacity), viscosity, swelling index and filtration loss.

Bentonite, the commercial name of montmorillonite from the clay minerals of the smectite group, shows colloidal properties when mixed with water, and its properties such as water swelling, high plasticity and ion exchange capacity are due to the three-layered crystal structure [21–28].

2. Sorption matter

The large surface used for industrial purposesnatural materials [7]. Absorbents and adsorbents generally used bentonite; Simectite, Atapulgite, Sepiolite. It can be classified as montmorillonite. The smectite group is one of clay minerals orkill more with more or less called bentonite. Bentonite base mineralmontmorillonite is common for the killer and is a commercially used term, at least soft, containing 85% montmorillonite, is an aluminum hydrosilicate with a colloidal property. When mixed with water, density of a few solid swelling bentonite about 2.5 g/cm^3. montmorillonite is calcium in many countries. Bentonite is a given name and the main contentwhich is montmorillonite and can change mainlycation can be defined as clay with Ca; Atapulgite, 2MgSi8O20 (H2O)4. The palygorskite expressed by the formula 4H2O an aqueous magnesium, aluminum silicate. Sepiolite is 6 Mg 9 Si 12 O 30(OH) 4 6H2O group is aqueous Mg silicate. In these mineralschannel-shaped poreswater bound to crystal structure with molecules. The clay in this group is micropore and channels and large surface areadue to the possession of various substancesabsorbers and adsorbing capacitiesIt is high.

2.1 Absorbent of bentonite and shale/clay features-waste water soil absorbent

Clay minerals in various industrial processes use, compositions and compositions are closely related. Grain size, grain shape, surface chemistry, surface ceramics, color, etching, viscosity, plasticity, absorption, adsorption v.b properties of clay minerals significantly impact on the use of. Absorption can be carried out in the presence of water or other liquid. Absorbents material is water and other liquids is a sponge as material containing zeolite and shale pores and the pores of the mass (solid material) such as shown in **Figures 6** and 7.

The main use of clay as an absorbent areas, ground absorbtion and cat litter, is the carrier of the drug. The high absorption capacity, the material large surface area, large pore volume, with sufficient pore size and distribution caused in developed matter of mass transfer and diffusion. In addition, its mechanical strength must increase when it gets wet. Bentonite, known as an absorbent clay, shale, sepiolite and atapulgite features have a large size absorbent material. Bentonite clay absorbance capacity, porosity, specific surface area, specific pore volume, and the pore size distribution of the acid, base and salt as well as chemical processes such as can also be increased by heat treatment [11–14]. Different absorption depending on clay type processes, for example, montmorillonite to the outer surface of the swollen hair water between the inner layers causing swelling which sorp waste water

Figure 6.
The composite sorbent use, zeolite distribution in pellet.

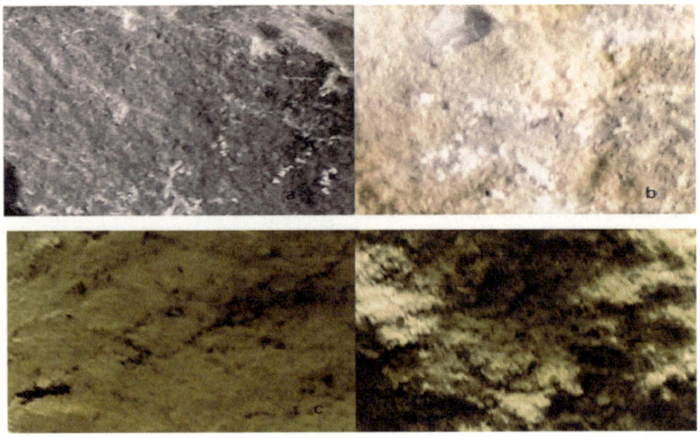

Figure 7.
The micro pictures Şırnak marly shale char shale as sorbent.

contamination. Sepiolite and atapulgite water absorption in a chain structure outer surfaces and zeolitic channels. In this type of structure of caged crystals there are no swelling between. The feature of the absorptive fluid is that the clay granules affect the absorption capacity. Liquid density, viscosity and surface capillary absorption of tensile clay granules are important factors affecting.

The absorbant clay bentonite was sorptive, colloidal, catalytic and rheological properties as given in **Table 2** [15]. Work in industrial waste watercleaning areas, non-burning, slippery to create a safe working environment absorbant killer is used [16–18, 28]. In the amount of clay used for hazardous waste water treatment increased over more than 180,000 tons/year.

2.2 Zeolite/asphaltite shale char composite

This area is especially montmorillonite, bentonite type clay used waste water treatmnent. The floor of the bentonite granules use as absorbant in 2020s started, but until the 2nd world war did not show improvement. The industrial waste

Unye bentonite	800–980 kg/cm^3
Sepiolite	400–700
Zolite	700

Table 2.
Shale and Marly shale bentonite properties.

waters: soma factory, airplane hangars, ship buildingbenches, other production facilities andIn the workshops, grease, oil, water, chemicals and other undesirable substances absorbed and cleaned.

Zeoite as filler material is commonly used as cat litter, granule powder do not build up, do not spread bad smell, granule grain size, basic as absorption capacity by the cat should be accepted. High absorption capacity having clay, only to absorb the urea not ventilated, but bad reduce smell and bacteria should avoid. The grain size distribution of clay granules it is important that it is usually between 1 and 6 mm is required. The beads are in the cat's claws. and its flanks and rounded surfaces should be. Cat litter, transport and use so as not to create dust during must have mechanical stamina [21].

2.3 Microwave treated briqutting of biomass char/zeolite composite

Washing of hazardous waste waters by microwave action efficiencies exceeding the total Fe Pb and Hg contents of sludges increased fast on coal char and wood char were also reported by Tosun [13].

2.3.1 Physical surface properties of char

BET specific surface areas, total surface activity, oxygen functional groups, total surface impurities, metal concentrations, dielectric value, free radical concentration and reactivity were related to the stimulation of oxidation reactivity. However, in some investigations, the pore size distribution of activated carbon is alsa likely to affect desorption kinetics [22–26].

3. Material and methods

3.1 Zeolite/carbon compost washing technology - sorbent applications

The chemical compositions of used local rock materials in waste sludge treatment (**Table 3**) [23].

During the experimental studies bentonite and zeolite samples, Ünye region, was investigated with intermediate type bentonite; pure, purified, tap water and

% Sorbent	Şırnak limestone	Sımak marly shalestone	Şırnak marl	Şırnak claystone	Expanded clay/zeolite
SiO_2	3,53	9.42	24,14	48,53	48,53
Al203	2,23	6,53	12,61	24,61	24,61
Fe2O3	0,59	4.48	7,34	7,59	7,59
CaO	49,48	39,23	29,18	9.48	9,48
MgO	2,20	2,28	4,68	3,28	9,28
K_2O	0,41	0,53	3,32	2,51	2,51
Na_2O	0,35	0,24	1,11	0,35	0,35
Ignition Loss	46,19	26,11	21.43	6,09	0,09
S03	0,32	0,21	0,20	0,32	

Table 3.
The chemical analvsis values of limestone, marly shale stone and clavstone of Şırnak province.

$CaCl_2.2H_2O$, NaCl, $MgCl_2$, KCl, $AlCl_3$ at concentrations ranging from 31,125 to 1000 ppm. Bentonite suspensions prepared by adding synthetic waters such as $6H_2O$ and bentonite suspensions were decanted by sedimentation method for 30 minutes in a 2 lt scale and bentonite concentrates were obtained and then necessary test and characterization procedures were applied afterwards.

Decantation was carried out in 2000 ml mills by adding 75 gr bentonite to 1900 ml of water. For a homogeneous suspension mortar, the bentonite water mixture was first subjected to scrub treatment in a Denver flotation cell for 5 minutes.

After the scurvy, the suspension was allowed to stand for 30 minutes after being agitated so that the impurities were precipitated. At the end of the period, suspended bentonite concentrate was removed by titration method and etch was dried.

The same procedure was repeated with synthetic waters prepared by adding salts at concentrations ranging from 31,125 ppm to 1000 ppm, until the bentonite concentrates were obtained in sufficient quantities with pure water, tap water and purified water.

The layout of the washing cycle is somewhat simpler than that of the lime slurry: there was no water–compost washing column towers connected to the waste sludge, and the washing unit contained one single microwave radiation column can be used to perform the three decantation washing phases: roughing, scraping and cleaning. The variation of the third cycle washing was also more limited recycled by microwave act.

The simple production presented as adapted and optimized depending on the target application. The main applications are briefly described in the following sections. Although this review only focuses on state-of-the art commercially available pellet plants, it should be noted that some prospective advanced applications for heat melting of binder are currently being studied, mainly in the form of prototypes or proof-of-concepts. These innovative applications include:

- Compost systems, in which the extrusion mold system takes advantage of temperature gradients in wet gradient.

- Compression press systems, where the high load press is used to drive the forming sludge in plant.

- Continous conversion systems, utilizing the high temperature binding gradients and amounts (of at least 20°C) in slurries to drive a recycle.

- Hot production, where the scraping power of the load system is used to drive the compressive form of hot system.

4. Results and discussions

4.1 Langmuir absorption model

For an overview of these more innovative and prospective applications, the general common method can be given in **Tables 4** and **5**.

The first order sorption concentration at three stage cycling counted bt the eqution below:

$$lnc^{Pb} = 1 + \frac{k_1 t}{1!} + \frac{k_1 t^2}{2!} + \frac{k_1 t^3}{3!}, 3ppm < x < 300ppm \tag{11}$$

Run	C, mg/l	k_1	a	b
1	28	0,3	0,15	1,2
2	20	0,24	0,22	1,7
3	12	0,21	0,27	2,4

Table 4.
The activated bentonite compostwith char shale of Şırnak materials.

Run	C, mg/l	k_1	a	b
1	28	0,3	0,15	1,2
2	20	0,24	0,22	1,7
3	12	0,21	0,27	2,4

Table 5.
The activated zeolite compost with char shale of Şırnak materials.

It has been found that the amounts of CaO and Na2O decrease due to the replacement of the Na^+ and Ca^+ releasable cations between the inde layers with the H^+ ions. Mg and Fe atoms in the octahedral crystal grains and the Al atoms in the octahedral centers, as well as the Al atoms in the tetrahedral layer, as well as Al2O3, MgO and Fe2O3. Make octahedral after bentonite from X-rays data elemental analysis data suggests that even lower coordination Al atoms are present in bentonite and less in the activated clay. The activated bentonite suspensions captured high level heavy metals such as Pb and Hg in the sludge (**Figure 8**).

Samples for this heat treated at different temperatures certain properties of the zeolite material (high absorbance capacity and not dispersed in the wet state) possibility of research and results as given in **Figure 9**.

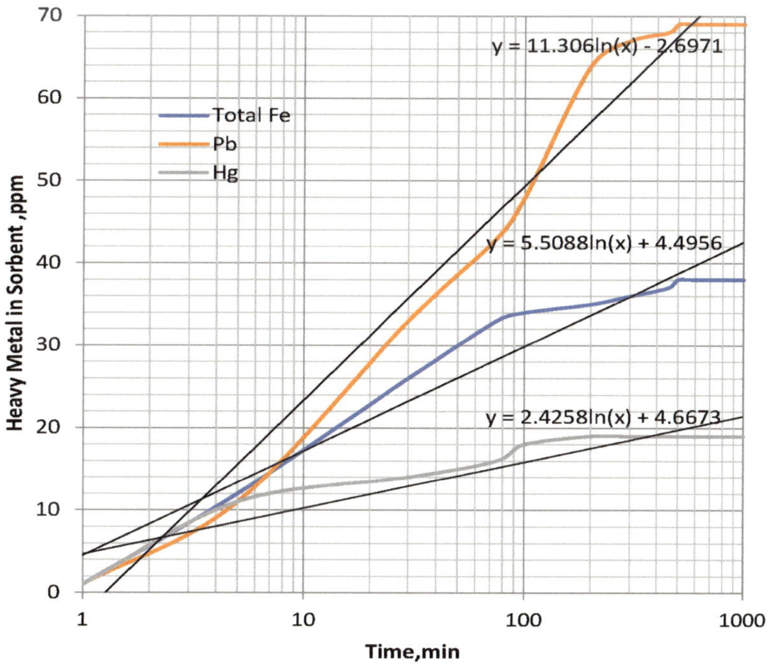

Figure 8.
The change in metal sorption depending on the metal concentration incorporated in the bentonite suspensions.

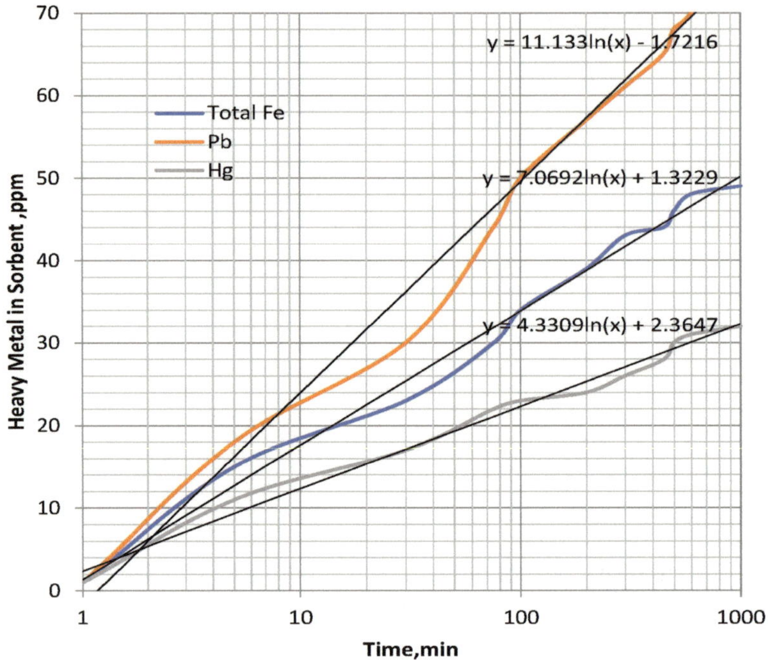

Figure 9.
The change in metal sorption depending on the metal concentration incorporated in the zeolite suspensions.

Cation exchange ability was so effective in metal sorption manner. The pH was efficient criteria in the washin column sorption.

It can be seen in the above graph, the pH decreases inversely proportional to the amount of salt added to bentonite suspension, which is much more noticeable when FeCl3 is used.

Bentonite is known to have a considerable dependence on the layer charge and edge charge pH. Therefore, a decrease in the cation exchange capacity should be expected in parallel with the decrease in pH.

The FeCl3 20 mg added bentonite solutions showed the change in cation exchange capacity (CEC, milliequivalent gram/100 gr) found in the bentonite concentrates and suspensions obtained using the precipitation-siphoning technique, depending on the salt concentration added.

4.2 Activation by HCl washing following methanol decay treatment under microwave radiation

The bentonite sample used in the study was obtained from Unye Madencilik from the Unye region of Tavkutlu mine. The bentonite sample was sieved and a small part of 45 μm was used for the operation. Bentonite samples were activated with 1 and 2 M HCl solutions for 2 h at 90° C using the Batch method (using 100 ml acid solution for 5 g sample). The acid-treated samples were washed with hot deionized water to remove Cl-ions and dried in room condition.

The microwave activated bentonite used in the experimental work was provided from the district of Unye in Ordu province. For the first time, Ünye region bentonite 0.1 M 100 ml CaCl2 solutions were mixed in the beaker at room temperature for 24 hours and the filtrate was converted to the ion-exchange by applying the AgNO3 test. Acid/clay suspensions were then prepared with bentonite, which was made to be ionized, to give H2SO4/clay ratios of 2 M, These were named -bentonite. These suspensions were dried at 150 oC for 3.5 hours. 50 ml of distilled water was added to the hot dry samples. The filtrate was filtered and dried at 80°C. Finally, 5 ml of chlorite was added to the acid activated bentonite samples, each of which was 0.01 g, and the shale was retained by exposure to methanol vapor at 60° C for 4 hours. In addition, these samples were further dried at the same temperature for 1 hour to remove weak chlorite species [22]. X-ray powder diffraction patterns of amorphous and acid activated bentonite samples were λ = 1.54050 Å wavelength Cu Kα/40 kV/40 mA RIGAKU 2200 diffractometer. Elemental analysis of the samples was performed using the ZSX 100e wavelength-separated X-ray fluorescence spectrometer (WDXRF) system with the Rigaku brand Rh anodic x-ray tube and the Rigaku SQX software package program. The specimens were stimulated with Rh anodic x-ray tube at 50 kV and 50 mA in order to reduce the damage that could occur in the samples during the measurement. PRIS Diamond brand TG/DTA thermal analyzer was used to obtain the thermal analysis curves of bentonite samples attached to pyridine. The thermal analysis curves of the samples were placed at a heating rate of 10°C/min, 5–10 mg of sample was placed in Pt crucible and taken against α-Al2O3 reference sintered at 20–700°C in air atmosphere. Again ATR spectra of the chlorite-bentonite samples were recorded in the absence of disc preparation with KBr under vacuum at a Bruker Vertex 80 V spectrometer at 1700–1350 cm $^{-1}$. The surface areas of raw, activated and shale clay samples were measured with a Quanta Chromosorb surface analyzer. The surface area was determined by measuring the thermal conductivity using a gas mixture prepared in 30% N2 and 70% He composition and taking into account the BET equation.

Thermal Properties were determined PRIS Diamond brand TG/DTA thermal analyzer. Thermogravimetric (TG) and Differential Thermal (DTA) analyzes of the samples used in the experimentation were carried out for dehydratation and activitation ability in sorption as seen in **Figures 10** and **11**.

4.3 Sorbent preparation for heavy metal absorption

The absorbance capacity of the microwave heated shale samples according to output zeolite, char shale and Ca-Bentonite experiments, the absorbance capacities were high, but high clay samples in water muddy have efficiently sorped Fe cations much. The results show that Na-Bentonite for this purpose seems unsuitable; from Breakup heat at high temperatures it must be processed but this the temperature of the absorbance has been detected. For this purpose the best result is about 400°C for Ca-Bentonite; the result is obtained at 200°C.

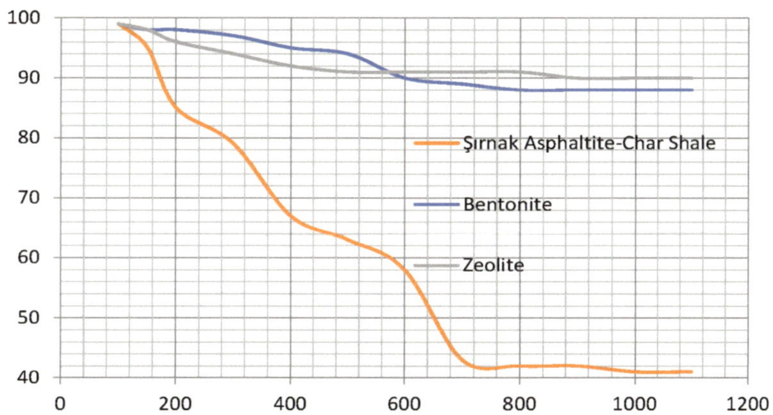

Figure 10.
TGA analyzes of the samples were carried out for dehydratation and activitation ability in sorption.

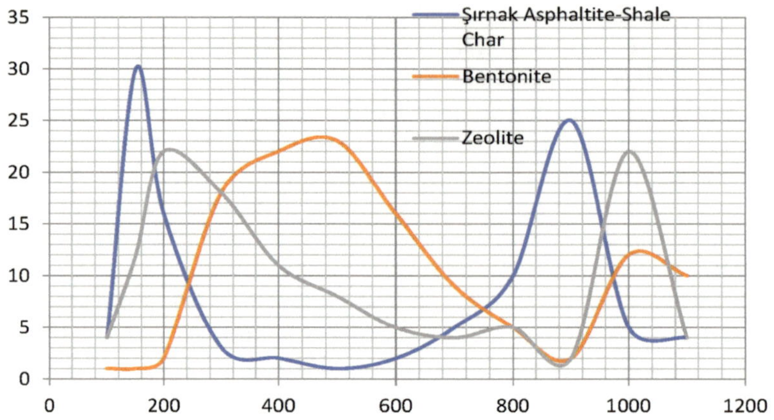

Figure 11.
DT analyzes of the samples were carried out for dehydratation and activitation ability in sorption.

The current use of absorbent clay and new areas of use increase in demand due to outflow. Especially absorbent clay market, the cat and cat market A significant improvement in America It was. For absorbent clay deposits our wealthy country, too, to have a significant share of there is no reason why. This absorbent is limited to meet clay consumption. The number of welding to, Turkey clays alternative can create resources. For this purpose, existing in Turkey absorbant clay beds must be fully identified, potential sources should be determined, absorbant purposefulness should be investigated and suitable properties absorbent clay production. Detailed studies to improve processesIt should be done. This study result, Turkey absorbanceon the samples taken from the By applying the processes, Afnor standardssuitable industrial absorbant clay productionIt was possible. Turkey's evaluating the existing potential to take part in this market, the country economywill provide significant benefits in terms of Compliance with environmental norms.

4.4 Compost of zeolite/shale/carbon pellets for heavy metal controls in waste waters

Bentonite is one of the clay minerals containing montmorillonite. Structurally, montmorillonite has a 2:1 layer of alumina octahedral (O) layer between two silica tetrahedral (T) layers [18]. The negatively charged excess coming from the isomorphic shifts is compensated by the interchangeable cations in the layers [18, 19, 28].

The acid activation process is widely used to improve the adsorption and catalytic properties of natural bentonites. The impurities, such as calcite and dolomite, are removed from the structure by the treatment of montmorillonite with inorganic acids, the interchangeable cations are replaced by hydrogen ions, and some of the Al ions in the tetrahedral layer dissolve certain cations of Fe, Al and Mg in the octahedral layer [19].

As a result, acid activation increases the pore diameters of the bentonite surface and the surface area and adsorption capacity up to a certain amount of this application [20]. If the amount of acid used during the acid activation process is excessively high, the Al ions found in the octahedral layer dissolve more and as a result, the mineral structure collapses, leaving a skeleton structure composed of silica solids. This reduces the adsorption capacity of the clay and disrupts its selectivity.

Pb is a colorless and Hg. The main sources are fossil fuels such as Pb, acidic mine waters and toxic metal sludges, which are industrial plants and industrial steel washings. [21] during the metal smelting processes and other industrial processes.

Pb compounds, one of the most common Hg pollutants, combine with water droplets within sulfuric acid aliquates or acid rain. Acidic mibe waters such as acidic rainfalls affect the chemical structure and biological conditions of the lakes and soil [22]. In addition, Pb Hg emissions are extremely harmful to health and directly affect the eyes, throat and respiratory tract [23].

Although the changes in the structural properties of bentonites after acid activation have been studied extensively in the literature, the studies on Pb adsorption of these samples are rather limited [24]. For this reason, the aim of this study is to investigate the thermal Pb and Hg washed adsorption properties of bentonites after acid modification and microwave activation.

The compost clays has substantial oxidation resistance and forming way service. This zero-pressure fluid provides precise, uniform temperature control to 500°C in closed-loop microwave systems where the heat transfer fluid is more than occasionally exposed to air. The fluid is comprised of a unique high-stability base plus high-performance oxidation inhibitor/stabilizer.

4.5 Carbon surface activation

In the sorbent size distrşbution, 80% of weights of samples were under 3 mm. The lignite samples were mainly distributed between 1 mm and 3 mm size fractions. The effect of particle size of solid sorbents were investigated over the combustion of Şırnak Asphaltite char shale and bentonite carried out well on acidic mine water of copper mine in Siirt substance subjected to reaction with bentonite clay in sorption, as shown in **Figure 12**.

Although metal diffusion on sorbent from waste sludge was befieved to be the primary mass transport process in the absorption chamber, complex reactions proliferated the alkali clusters below 1-2 mm size and exothermic oxidation reactions increased toxic substances in the effluent form, a relatively porous structure of bentonite clay interstitial spaces and cracks reduced below 1 mm size. The hazardous heavy metal concentrations reacted adsorbate then adsorbs to the sorbent in an certain amount that is equal to the amount of previous adsorbate that was partially degraded on the surface of the bentonite clay and stuck covered toxins, along with avoiding chelating organic matter related carbonil and amine.

4.6 Washing sorption by clay compost control

The waste water washing provide the main support to the clean water production. The commercial successes in clay mud mentioned in washed bed and its

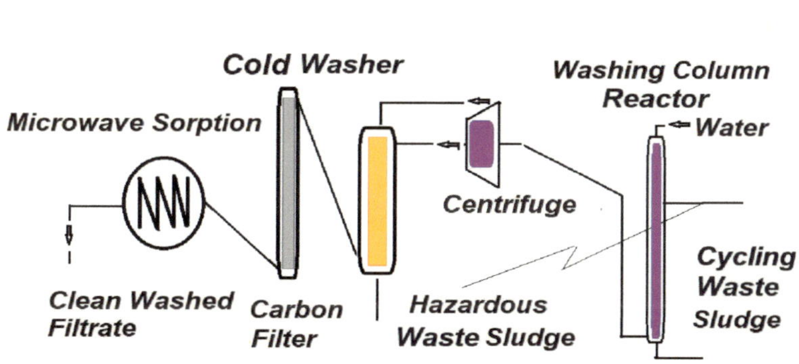

Figure 12.
Schematic view of an washing with microwave recycled by microwave sorption technique.

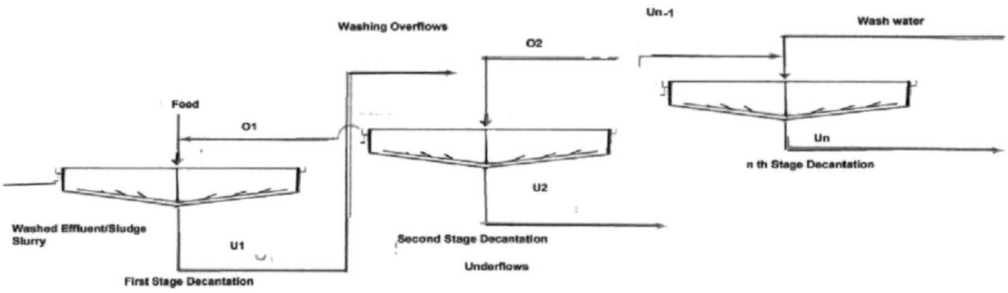

Figure 13.
The washing decantation by sorbent use, leaching in microwave activation as sorbent compost.

sedimentation ability were described some of the emerging applications in lime use like clean water neutralization, aeration [26]. The figures of statistical potential of washing control with different techniques in waste water cleaning are classified as seen in **Figure 13** regarding cycling decantation time.

Some cost evaluations covering security of supply and environmental impacts, climate change evaluations, and technical and economic analysis, may be disussed in cycling cost and activities [29–32].

Initially, most of the toxin removal occurs through chemical adsorption of the toxins to the expanded clay where the combustion temperature was in the combustion phase below 750°C that lasts approximately 2–3 mins. The removal efficiency of 40–90% were reported during this temperature range. Total organic toxin substances were completely slightly at efficiencies of 75–90% in the late combustion phase. The pictures of zeolite fines soaked in char shale/clay were illustrated in **Figures 9** and **10**. Following the combustion at 800°C adsorbed film of emissions over expanded clay was shown in **Figure 13**.

A common industrial combustion to control the emissions pro combustion stage lime washing involves backwashing with air and hydrated lime water rinse. Process variables include the control backwash rate, surface wash rate/duration, time sequence and duration of backwash. Clean filtrate is pumped back into the bottom of the column during backwashing.

5. Waste water treatment

The required test and characterization procedures such as pH viscosity measurement, filtration loss and swelling index were applied to all bentonite concentrates obtained and then to products activated with 0.5% soda.

These sorbents need to be accurately mixed with combustion matter and to optimize the combustion process. Reliable models, based on the above results, need ta be combustion chamber construction far the estimation of kinetic parameters for toxic stream control. Such toxic stream circulation models would aid in the microwave activated shale clay sorbent use in waste water treatment systems as shown in **Figure 8**.

The country needs the cleanest fuel to be produced providing the essential oils and gases. For this reason, acidic mine waters as heavy metal contamination to control fish farming were mixed with expanded clay at 1–2 mm size soaked with slurries of different alkali sorbents such as bentonite, shale fine, NaCl, CaCl2 and KCl were tested in the packed bed column washing and the test results were illustrated in **Figure 10**.

A decantation bed thickner was used in compost sorption process was tested at a scale of 2–3 kg/h; collecting operational and design data to build an industrial installation. A technological diagram of the compost washing at three stage process developed unit was made. Activated shale destruction almost observed at third cycled end. Heavy metal concentration change increased from 2nd cycle with performance of 60–70% and also simultaneous dilution of waste mud products by sedimented. it is necessary to optimize the cycling stages on metal circulation without the metal concentration change.

5.1 Sorbent composite char/salt method

These cheap alkali sorbent fines may be so feasible at the side of cost and sorbent production. The high amount toxicity of hazardous waste slurries of cyanide in recovery of Au and hot water streams might be reduced by massive alkali sorbent use [13]. Advanced column washing by char of Turkish lignite and wood char may be feasible. However, the washing with solid expanded clay soaked alkali fine char of Turkish lignites can be utilized [14]. The heavy metal and toxic contaminants of waste waters, nitrates nitrites pesticides and Hg with clay soaked at high elimination rates ranging 52–64%.

The approach of sorption kinetics assumed basically that the process exponentially was developed itself, as seen in **Figure 8** with specific features. The elimination of Pb and Hg in waste waters with clay compost sorbent was a decisive sorbent for the reaction path on the kinetics of washing acidic mine waters. Therefore a static model of washing decantation was developed at 1 mm of coal sand size. Lnstead of fluid bed combustion, packed bed column of coarse size composts and

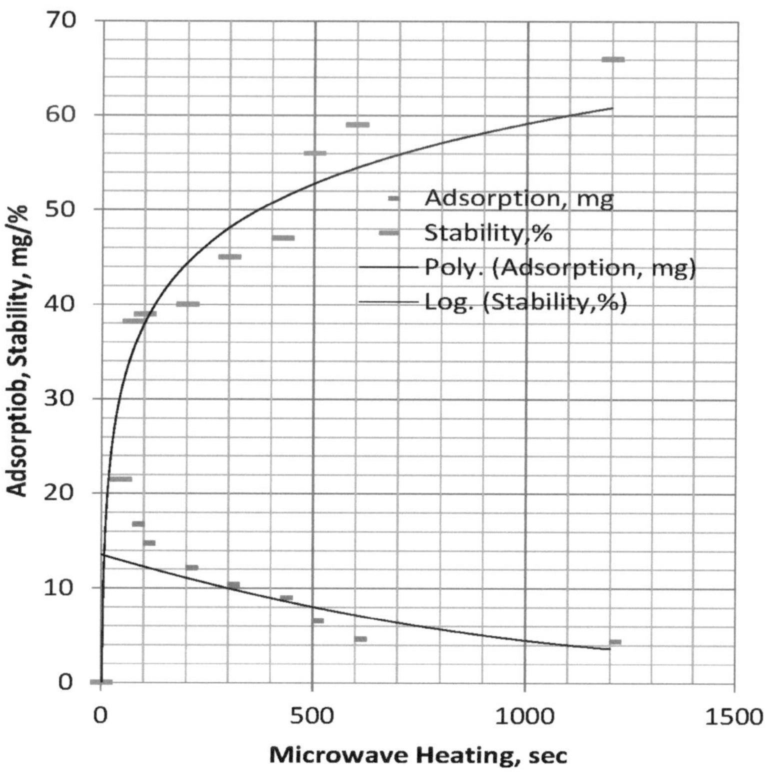

Figure 14.
Thermal stability of raw and activated bentonites with chlorite of char shale.

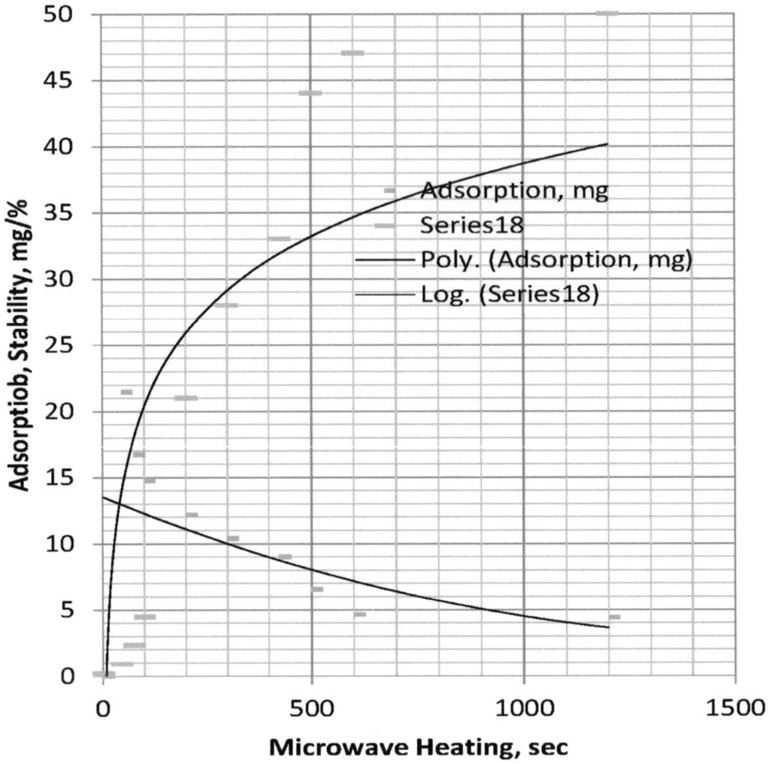

Figure 15.
Thermal sorption quality and activated zeolite with chlorite of char shale.

packages wass highly governed by slow washing, sufficiently sorbtion of toxic sludges.

As seen in **Figure 13**, the activated clay - zeolite examined was more efficient as an absorbent for a conversion of waste waters to friendly watres. It can be a promising waste contaminated waters, municipal wastes because of high activity in the collection and leaching in the toxic solutions in the mine waters and lakes. The sorption rates reached to 74% with Şımak asphaltite shale and zeolite compost.

However, due to the presence of salt in the environment, the bentonite particles which have become dispersed and suspended must coagulate and collapse along with other large nasties, as the surface loads decrease to absolute value and approach zero load point. The high decreases in the separation efficiency in the FeCl3 medium observed in **Figure 14**.

The activation decreased in ambient pH. So, the most important ions in the bentonites are H + and OH. Therefore, the changes that may occur in pH especially at low pH H + ion adsorbed the bentonite particles to the surface and makes the surface neutral to facilitate coagulation, so the bentonite particles which are to be suspended in the precipitation conditions are also collapsed and the separation efficiency is degraded at low pH. It has been reflected as seen in **Figures 14** and **15**.

The ability of the Na^{+} ion to hydrate and especially the Ca^{+2} cation to take its place easily is an important consideration.

6. Conclusions

The corresponding increase in surface area can iöprove the removal of impurities in the sample due to the microwave dissolved acid treatment over zeolite/

asphaltite char shale and bentonite applied and by the replacement of the interchangeable cations by the H+ cation. The increase in the surface area of the bentonite type after acid modification heay metal sorption 54%.

As the acid molarity increased, it was determined that the acid-modified Bentonite (0.84 ppm) and Zeolite (0.69 ppm) samples adsorbed more Pb than the natural form B (0.66 ppm).

Due to the removal of octahedral cations after acid activation, the formation of new acid sites in the clay lamella structure increased the specific surface area and porosity. Therefore, the structure of bentonite showed more Pb and Fe adsorption properties. In this study, the Pb adsorption capacity of bentonite samples prepared with 100 min circulation was found to be the sufficient washing.

Analyzes were performed using approximately 30 mg of sample at a temperature range of 30–1000°C at a heating rate of 10°C/min. The temperature ranges and mass loss values obtained from the TG analysis are given in **Table 6**. The endothermic peaks observed in the DTA curves of the natural Bentonite, char shale and zeolite samples at 105, 299 and 98°C, respectively. There are due to the removal of physically adsorbed water.

The shift of endothermic peak temperatures to lower values with increasing acid concentration was also observed. TGA analysis of natural bentonite, coal char shale and zeolite samples revealed that the total mass losses at 1000°C were 9.19%, 47.75% and 8.15%, respectively.

This high divergence was the endothermic peak at 150°C in DTA. In addition, in the TGA curve, a mass loss of 4.3% at 200–320° C is the endothermic peak at 270°C in DTA. This mass loss is due to the removal of hydrate and sulphate bound to Brønsted acid centers. In addition, mass loss of 4.9% in the range of 390–650°C due to dehydroxylation of the crystal lamella layers is in the form swelling easily on an endothermic peak of 621°C centered on DTA.

In the pH measurements made, the pH value of 7,3 in washing hazardous waste water finally at the last washing column decreased to 5, depending on the concentration of salt content of sorbents in the water.

In the thre stage microwave activated washing test measurements made with tap water, it was found that 73 mg/l (ppm) in bentonite/asphaltite char shale decreased to 53 mg/l (ppm)/in last column output. Likewise, the washed waste waters obtained after 100 min washing by microwave activaty using sodium salts softed flow with 1 mm sorbent packages showed reductions in Pb, Hg and Fe at 47% performance.

In clean water aliquate had the 24 ppm Pb,5 ppm Hg and 57 Fe values, which Pb reduction rates of sorption at Langmuir model with nitrate washed, was 0,73 ppm/min.l, Hg and total Fe reduction rate has decreased to 0,43 ppm/min.l and 0,23 ppm/min.l, respectively.

	SiO_2	Al_2O_2	Fe_2O_2	MgO	K_2O	CaO	TiO_2	LOI*
Kaolin (%)	47.85	37.60	0.83	0.17	0.97	0.57	0.74	11.27
Şırnak Asphaltite Char Shale	47.85	37.60	0.83	0.17	0.97	0.57	0.74	11.27
Bentonite	47.85	37.60	0.83	0.17	0.97	0.57	0.74	11.27
Marly Shale	47.85	37.60	0.83	0.17	0.97	0.57	0.74	11.27
Zeolite	47.85	37.60	0.83	0.17	0.97	0.57	0.74	11.27

**LOI: Loss on Ignition at 1000°.*

Table 6.
Sorbent Clay Types for waste water treatment.

The swelling index and viscosity studies of compost bentonite and zeolite did not changed in packed columns washing filtered through steel meshed packages during microwave act washing.

The pH increased at washing was efficient in heavy metal sorption, the swelling index decreased, the loss of filtration increased negatively, and viscosity decreased by the addition of sodium.

In the obtained data, it was observed that sorption manner of bentonite has negatively effected by foreign ions in washing water for the activation especially total iron ion.

This result also indicated that the properties of the irrigation and fish farming water to be used during wet soil amendment of of agricultural organic soil and lake muds with wet bentonites, on waste water treatment units which friendly mud should be controlled, otherwise the contamination after discharge would be harm human health, toxicology of animal and fish feed.

Abbreviations

Greek symbols

a	affinity parameter of the Langmuir isotherm (L mg^{-1})
b	stoichiometric constant defined by
B	reactant solid defined
Bi_m	Biot number for mass transfer
C_i	concentration of manganese in the bulk external phase of stage i (mg L^{-1})
C_0	feed concentration of manganese in the column (mg L^{-1})
D_{ef}	effective diffusion coefficient (m^2 s^{-1})
F	objective function
h	fixed bed height (m)
k_e	mass transfer coefficient in the bulk external phase (m s^{-1})
k_r	reaction rate constant for heterogeneous systems (m s^{-1})
N	number of stages
Q	volumetric flowrate (m^3 s^{-1})
q_i	concentration of immobilized manganese within the adsorbent particle at stage i (mg g^{-1})
q_m	theoretical maximum adsorption capacity of the Langmuir isotherm (mg g^{-1})
r	radial distance from the center of the particle, $0 < r < R_p$ (m)
R	radius of column (m)
R_p	radius of adsorbent particle (m)
R^2	determination coefficient (–)
$r_{c,i}$	unreacted core radius at stage i (m)
t	time (s)
V_i	volume of stage i (L)
α	backmixing coefficient (–)
φ	column hold-up (–)
ρ	density of adsorbent particle (g m^{-3})
τ	mean residence time of fluid in the column (s)

DOI: http://dx.doi.org/10.5772/intechopen.94404

Author details

Yıldırım İsmail Tosun
Mining Engineering Department, Engineering Faculty, Şırnak University, Şırnak, Turkey

*Address all correspondence to: yildirimismailtosun@gmail.com

References

[1] Nalbantcilar, MT, Pinarkara, SY, 2016, Public health risk assessment of groundwater contamination in Batman, Turkey, Water Health. 2016 Aug;**14**(4): 650-61. doi: 10.2166/wh.2016.290.

[2] Çelik, R., 2015, Temporal changes in the groundwater level in the Upper Tigris Basin, Turkey, determined by a GIS technique, Journal of African Earth Sciences, Volume 107, July 2015, Pages 134–143

[3] Disli, E., 2017, Hydrochemical characteristics of surface and groundwater and suitability for drinking and agricultural use in the Upper Tigris River Basin, Diyarbakir—Batman, Turkey, Springer, 2017, Environmental Earth Sciences **76**(14), DOI: 10.1007/s12665-017-6820-5

[4] S. Ahamed, A. Hussam, A.K.M. Munir, 2009, Groundwater Arsenic Removal Technologies Based on Sorbents: Field Applications and Sustainability, Handbook of Water Purity and Quality, Academic Press, Amsterdam (2009) 379–417

[5] J.S. Ahn, C.M. Chon, H.S. Moon, K. W. Kim, 2003, Arsenic removal using steel manufacturing by-products as permeable reactive materials in mine tailing containment systems, Water Research, **37** (2003), pp. 2478–2488

[6] J.P. Allen, I.G. Torres, 1991, Physical separation techniques for contaminated sediment, N.N. Li (Ed.), Recent Developments in Separation Science, CRC Press, West Palm Beach, FL (1991)

[7] S.J. Allen, L.J. Whitten, M. Murray, O. Duggan, 1997, The adsorption of pollutants by peat, lignite and activated chars, Journal of Chemical Technology & Biotechnology, **68** (1997), pp. 442–452

[8] E. Álvarez-Ayuso, H.W. Nugteren, 2005, Purification of chromium(VI) finishing wastewaters using calcined and uncalcined Mg-Al-CO3-hydrotalcite, Water Research, **39** (2005), pp. 2535–25

[9] R.A.I. Abou-Shanab, J.S. Angle, R.L. Chaney, 2006, Bacterial inoculants affecting nickel uptake by Alyssum murale from low, moderate and high Ni soils, Soil Biology and Biochemistry, **38** (2006), pp. 2882–2889

[10] R.A. Abramovitch, B.Z. Huang, M. Davis, L. Peters, 2003, In situ remediation of soils contaminated with toxic metal ions using microwave energy, Chemosphere, **53** (2003), pp. 1077–1085

[11] J. Acharya, J.N. Sahu, C.R. Mohanty, B.C. Meikap, 2009, Removal of lead(II) from wastewater by activated carbon developed from Tamarind wood by zinc chloride activation, Chemical Engineering Journal, **149** (2009), pp. 249–262

[12] J. Acharya, J.N. Sahu, B.K. Sahoo, C.R. Mohanty, B.C. Meikap, 2009, Removal of chromium(VI) from wastewater by activated carbon developed from Tamarind wood activated with zinc chloride, Chemical Engineering Journal, **150** (2009), pp. 25–39

[13] Novak, I., Cicel, B. 1978. Dissolution of smectites in hydrochloric acid; II, Dissolution rate as a function of crystallochemical composition. Clays and Clay Minerals, **26**, 341–344.

[14] Önal, M., Sarikaya, Y., Alemdaroglu, T., Bozdogan, I. 2002. The effect of acid activation on some physicochemical properties of a bentonite. Turkish Journal of Chemistry, **26**, 409–416

[15] Srasra, E., Bergaya, F., van Damme, H., Arguib, N.K. 1989. Surface properties of an activated bentonite. Decolorization of rape-seed oil. Applied Clay Science, **4**, 411–421.

[16] Srivastava, R.V. 2003. Controlling of SO2 Emissions, A Review of Technologies, Nova Science Publishers, Inc., New York, 1 pp.

[17] Venaruzzo, J.L., Volzone C., Rueda M.L., Ortiga J. 2002. Modified bentonitic clay minerals as adsorbents of CO, CO2 and SO2 gases. Microporous and Mesoporous Materials, **56**, 73–80.

[18] Alemdaroglu, T., Akkus, G., Önal, M., Sarikaya, Y. (2003). Investigation of the Surface Acidity of a Bentonite Modified by Acid Activation and Thermal Treatment. Turkish Journal of Chemistry, **27**, 675–681.

[19] Benesi, H.A. (1956). Acidity of Catalyst Surfaces I. Acid Strength from Colors of Adsorbed Indicators. The Journal of Physical Chemistry, **78**, 5490–5494.

[20] Benesi, H.A. (1957). Acidity of Catayst Surfaces II. Amine Titration Using Hammett Indicators. The Journal of Physical Chemistry, **61**, 970–973.

[21] Caglar, B., Afsin, B., Tabak, A. (2007). Benzamide Species Retained by DMSO Composites at a Kaolinite Surface. J. Therm. Anal. Cal., **87**, 429–432.

[22] Caglar, B., Afsin, B., Tabak, A., and Eren, E. (2009). Characterization of Cation Exchanged Bentonites by XRPD, ATR, DTA/TG and BET measurement Investigation of. Chemical Engineering Journal, **149**, 242–248.

[23] Chitnis, S.R., Sharma, M.M. (1997). Industrial Applications of Acid-Treated Clays as Catalyst. Reactive and Functional Polymers, **32**, 93–115.

[24] Christidis, G.E., Scott, P.W., Dunham, A.C. (1997). Acid Activation and Bleaching Capacity of Bentonites from the Islands of Milos and Chios, Aegean, Greece, Appl. Clay Sci., **12**, 329–347.

[25] Flessner, U., Jones, D.J., Roziere, J., Zajac, J., Storaro, L., Lenarda, M., Pavan, M., Lopez, A.J., Castellon, E.R., Trombetta, M., Busca, G. (2001). A Study of the Surface Acidity of Acid-Treated Montmorillonite Clay Catalyst. J. Mol. Catal. A: Chemical, **168**, 247–256.

[26] Heyding, R.D., Ironside, R., Norris, A.R., ve Prysiazniuk, R.Y. (1960). Acid Activation of Montmorillonite. Can. J. Chem., 38, 1003–1016.

[27] Hutson, N.D., Hoekstra, M.J., Yang, R.T. (1999). Control of Microporosity of Al2O3-Pillared Clays: Effect of pH, Calcination Temperature and Clay Cation Exchange Capacity. Micropor. Mesopor. Mater., 28, 447–459.

[28] Volzone, C., Ortiga, J. 2009. Adsorption of gaseous SO2 and structural changes of montmorillonite. Applied Clay Science, 44, 251–254

[29] Noyan H, Önal M, Sarıkaya Y. The Effect of Sulphuric Acid Activation on the Crystallinity, Surface Area, Porosity, Surface Acidty, and Bleaching Power of a Bentonite. Food Chemistry. 2007;**105**: 156-163

[30] Önal M, Sarıkaya Y, Alemdaroğlu T, Bozdoğan İ. The Effect of Acid Activation on Some Physicochemical Properties of a Bentonite. Turkish Journal of Chemistry. 2002;**26**:409-416

[31] Reedy, C.R., Nagendrappa, G., Prakash, B.S.J. (2007). Surface Acidity Study of Mn+-Montmorillonite Clay Catalysts by FT-IR Spectroscopy:

[32] Rodriguez MAV, Barrios MS, Gonzalez JDL, Munoz MAB. Acid Activation of a Ferrous Saponite (Griffithite): Physico-Chemical Characterization and Surface Area of the Products Obtained. Clays and Clay Minerals. 1994;**42**:724-730

Section 2

Applications

Chapter 5

Limestone Clays for Ceramic Industry

Herbet Alves de Oliveira and Cochiran Pereira dos Santos

Abstract

Limestone clays are used in the ceramic segment in the manufacture of bricks, ceramic tiles, and in the production of cement, among others. Limestone can be present in soils in pure form or as a contaminant, but always from marine environments. The limestone after burning can present a high loss of mass (35–45%), which can cause serious problems with the sintering of ceramic products such as bricks, tiles. The calcium or magnesium carbonate once dissociated forms calcium oxide (CaO) and releases carbon dioxide (CO_2). CaO in contact with water subsequently experiences very high expansions that can cause cracks in the materials. Researchers have studied procedures to inhibit limestone action on clays as well as to set the correct temperature for firing. In this chapter, examples of clays with different percentages of calcium carbonate ($CaCO_3$) that are used in the ceramic segment and their characteristics will be given.

Keywords: clays, ceramic, limestone clays, calcium carbonate

1. Introduction

Clays are inorganic, natural, earthy, and fine-grained materials that acquire plasticity when mixed with water [1]. For sedimentologists, a clay is a raw material whose grain size is less than 2 μm. Like clays, in turn, there are rocks made up of clay minerals and may contain other minerals such as quartz, feldspar, mica, calcite, hematite, and organic matter as accessories [2]. A clay, once ground and mixed with water, in addition to presenting excellent workability in the fresh state, after drying, becomes extremely rigid. After burning normally above 800°C, it acquires great resistance [3]. Clays are used worldwide in the ceramic industry, especially in bricks, coatings, and others. However, clays are formed from the weathering of explosion and can be contaminated with several minerals among them or carbonate, which can alter the shape that causes the following burns. Limestone may be present in colloidal form, or coarse particles. However, in all cases it is impossible to separate or calculate this. Some researchers have tried to reduce the size of the variations to improve the chemical changes. According to Barba et al. [4], calcium carbonate and magnesium carbonate are the main constituents of carbonate sedimentary rocks. Anionic carbonate groups are strongly activated units and share oxygen with each other. They are responsible for the properties of these minerals. The most important anhydrous carbonates belong to three isostructural groups: the calcite group, the aragonite group, and the dolomite group. Among these, the minerals most used in the ceramic industry are calcite and dolomite, as they are low-cost raw materials, in addition to having favorable physical and chemical properties and available deposits. Second, Padoa [5] adds that when $CaCO_3$ is small, a decomposition can be complete and the calcium oxide reaches later with other mass components forming calcium silicates and silicon aluminates (wollastonite, anortite, gehlenite etc.) during sintering. Barba et al. [4] mentioned that the raw materials of clay when burned at high temperatures produce crystal phases that influence the properties of ceramic products. Calcite exerts a bleaching action on burnt products when added to a formulated mass of clays (in proportions above 5% and less than 30%) and at the same time decreases its expansion by legislation, as it forms crystalline and liquid phases, including cycles temperature and firing adopted. Calcite and dolomite are the most important representatives of carbonates in the ceramic industry. They are used as main components in the manufacture of ceramic tiles with high water absorption. These coatings include "porous coatings" or "tiles." These products are designed or used on walls and are not suitable for application on floors, as they have undesirable technical characteristics, such as mechanical resistance, incompatibility with use. According to Amorós [6], properties of parts of a ceramic product are registered by crystalline phases formed based on calcium and magnesium as ghelenite ($SiO_2 \cdot Al_2O_3 \cdot 2CaO$) and anortite ($2SiO_2 \cdot Al_2O_3 \cdot CaO$). To achieve these phases, use the dolomite calcium oxide and/or magnesium reaction with a remaining clay structure proven by its thermal decomposition.

The calculation in general can affect the ceramic product in two ways: low percentages (up to 3%) and high temperature (above 1180°C) result in flow agents, that is, materials that contribute to reduce water absorption and increase the resistance of ceramic products. Above 3%, they can act as a foundation at temperatures above 1170°C [7].

In this chapter, we will highlight properties of limestone clays and their application in the ceramic industry.

2. Clays

Clays are hydrated aluminum silicates with crystalline structure arranged in layers, consisting of continuous sheets of SiO_4 tetrahedrons, ordered in a hexagonal shape, condensed with octahedral sheets of di and trivalent metal hydroxides, usually below 2 μm. They are materials that in contact with water become plastic, a fundamental characteristic for conformation of ceramic products because it provides mechanical resistance in the pressing, extrusion, or gluing process. Clays are mixtures of various clay minerals such as kaolinite, illite, and montmorillonite, which may or may not contain impurities [3, 8].

2.1 Kaolinite

The kaolinite with structural formula $Al_2O_3 \cdot 2SiO_2 \cdot 2H_2O$ has a dioctahedral structure, which consists of a tetrahedral layer linked by an octahedral layer. Pure kaolinites usually have low plasticity, see **Figure 1**.

2.2 Montmorillonite

Montmorillonites are a set of family of clay minerals, composed of dioctahedral and trioctahedral silicate sheets, see **Figure 2(a)** and **(b)**. The most outstanding feature of these minerals is their ability to absorb water molecules [8, 9]. It has 80% of exchangeable cations in the galleries and 20% on the lateral surfaces. The modification of montmorillonite clays has aroused scientific and technological interest

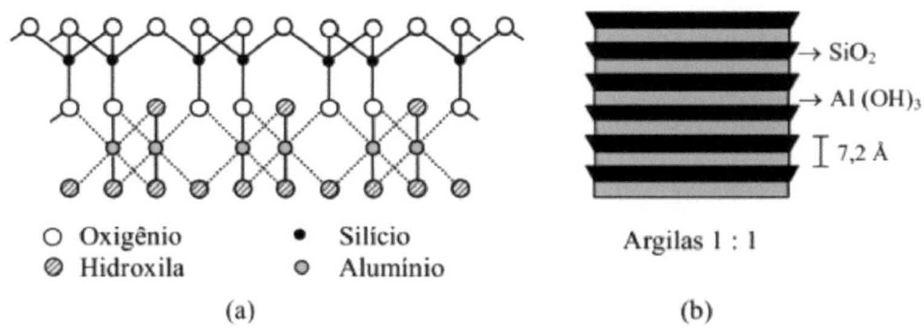

Figure 1.
Kaolinite structure. (a) Si—O tetrahedra on the bottom half of the layer and Al—O,OH octahedra on the top half. (b) Dioctahedral structure.

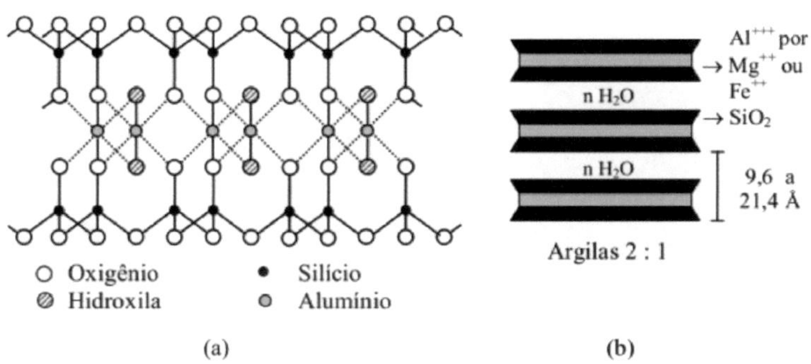

Figure 2.
Crystalline structure of a montmorillonite. (a) Montmorillonite structure, composed of Si, Al, and O. (b) Sheets of dioctahedral and trioctahedral silicates.

for providing significant improvements when incorporated into pure polymeric materials and conventional composites. The clay modification process occurs preferably through the ionic exchange of the exchangeable cations of its crystalline structure.

2.3 Illite

The basic structural unit of the illites is the same as that of the montmorillonites except that in illites, the silicon atoms in the silica layers are partially replaced by aluminum. Therefore, there are free valences in the boundary layers of the structural units, which are neutralized by K cations, arranged between the overlapping units. The structural scheme of the illites is shown in **Figure 3**. The K cation is the one that best adapts to the hexagonal meshes of the oxygen planes of the layers of silica tetrahedron and is not displaced by other cations. The water adsorption and cation exchange capacity is due only to the broken connections at the ends of the layers. The average diameter of the illites varies between 0.1 and 0.3 μm. When the replacement of silicon in the tetrahedron layers by aluminum in the illites is small, the connections between the structural units provided by the K cations may be deficient and will allow water to enter. When this occurs, the properties of the illites are close to the properties of montmorillonites [3].

2.4 Chlorite

Chlorites are minerals made up of four hydrated aluminum and magnesium silicate layers, containing Fe (II) and Fe (III) as shown in **Figure 4**.

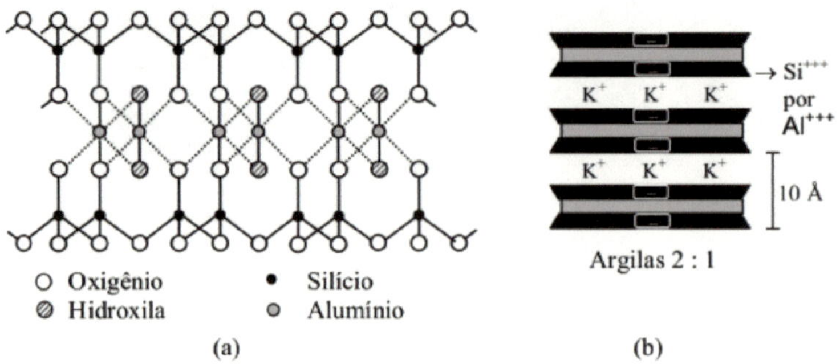

Figure 3.
Crystalline structure of an illite. (a) Silicon atoms in the silica layers partially replaced by aluminum in the illites. (b) Structural scheme of illites.

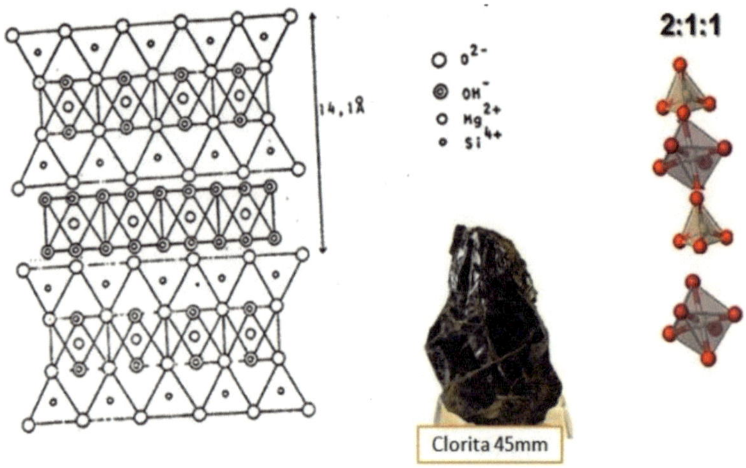

Figure 4.
Crystalline structure of chlorite [9].

The most common clay minerals are interstratified, characteristic of mixtures of clay minerals, classified by subgroup and mineralogical species, see most common classification in **Table 1**. Clay minerals are divided into several classes. A large majority of clays do not have in just one crystalline phase. Two or more chemical species may be present.

3. Clays used in the ceramic manufacturing process

The clays used in the ceramic manufacturing process can be classified into:

- Carbonitic clays: they are formed by associations of illitic-chloritic and eventually illitic-kaolinite clay minerals. The amount of calcium carbonate present can be variable. These clays give the dough plasticity. Generally, after burning they have colors ranging from beige to orange [4].

- Non-carbonitic clays: they are characterized by the almost total absence of carbonates. The clay minerals present are of the illitic-chloritic type. It has the function of giving plasticity to the dough, and generally after firing they give rise to well-sintered materials.

Subgroup	Chemical species	Minerals
Kaolin $X_n(Y_2O_5)(OH)_4$	Kaolinites	Nacrite ($Al_2(Si_2O_5)(OH)_4$) Dikite ($Al_2(Si_2O_5)(OH)_4$) Livesite ($Al_2(Si_2O_5)(OH)_4$) Halloysite ($Al_2(Si_2O_5)(OH)_4$)
Talc $X_B(Y_2O_5)(OH)_2Z_mH_2O$	Montmorillonites	Montmorillonites ($Al_{1,51}Fe_{0,07}Mg_{0,60})(Al_{0,28}Si_{3,72})O_{10}(OH)_2Na_{0,33}$
		Beidellite ($Al_{1,46}Fe_{0,50}Mg_{0,08})(Al_{0,36}Si_{3,64})O_{10}(OH)_2Na_{0,4}$
		Nontronite ($Fe_{1,67}Mg_{0,33})(Si_4O_{10})(OH)_2Na_{0,33}$ and $Fe_{2,22}(AlSi_3O_{10})(OH)_2Na_{0,33}$
		Hectorite ($Mg_{2,67}Li_{0,33})(Si_4O_{10})(F,OH)_2Na_{0,33}$
		Saponite $Mg_3(Al_{0,33}Si_{3,67})O_{10}(OH)_2Na_{0,33}$
	Illites	Wide variety of minerals
Chlorite	Chlorites	Chlorite
$X_{2n}(Y_2O_5)_2(OH)_2$		$[Mg_2(Al,Fe(III))(OH)_6][Mg_3(AlSi_3O_{10})(OH)_2]$

Source: Barba et al. [4].

Table 1.
Subgroups of clay minerals.

- White plastic clays: the clay matrix is kaolinitic, with little illite. They give plasticity to the dough, and after burning they have a white color.

- Kaolinitic clays: clays of low plasticity and normally free of fluxing oxides such as K_2O and Na_2O, therefore, with refractory characteristics.

3.1 Heat action on clays

According to Mackenzie [10], when a ceramic raw material is subjected to the action of heat, it experiences volumetric variations, usually permanent and irreversible, which can be classified as:

- Oxidation of organic matter

- Decomposition of compounds containing oxygen, such as sulfates, carbonates, etc.

- Dehydroxylation of the clayey mineral

- Crystallization by increasing the temperature

- Vitreous phase formation

- Solid solutions: adjacent crystals of two different materials but of similar structure can react with each other, forming a solid solution.

Kaolinitic clay: the scheme according to **Figure 5** shows an endothermic peak between 560 and 590°C referring to the elimination of hydroxyls from the constitution water present in the clays, and an exothermic peak between 980 and

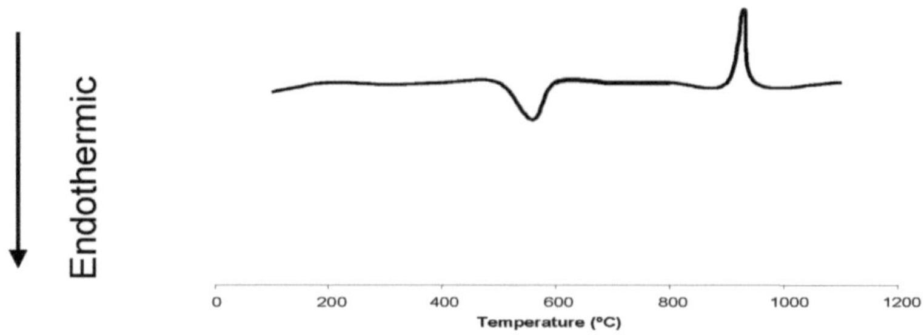

Figure 5.
Differential thermal analysis of a kaolinitic clay [10].

Figure 6.
Differential thermal analysis of a montmorillonite clay [10].

1000°C, due to the formation of mullite, which can be represented by the reactions 1 and 2 [8].

$$Al_2O_3.2SiO_2.2H_2O \xrightarrow{580\text{-}600} Al_2O_3.2SiO_2 + 2\ H_2O \quad (1)$$

$$3(Al_2O_3.2SiO_2) \xrightarrow{980\text{-}1000} 3Al_2O_3.2SiO_2 + 4SiO_2 \quad (2)$$

Montmorillonite: montmorillonites have water that lodges in the mineral structure, that is, hydration water of adsorbed ions. The elimination of hydroxyl groups occurs at 700°C. At 850°C, a small endothermic peak may occur due to the loss of montmorillonite crystallinity. Illites can present loss of adsorbed water between 100 and 200°C and water loss in the constitution between 550 and 600°C, see **Figure 6**.

3.2 Most common impurities present in clays

Quartz: it appears in clays in colored or colorless round grains, whose percentage ranges from 0 to 60%. For high levels of quartz, the clay is called sandy and has low plasticity [11].

Hematite: iron can be present in the forms of hematite (α-Fe_2O_3), goethite (α-$FeO{\cdot}OH$), and lemonade (a mixture of iron oxides and hydroxides of a weakly crystalline nature), or simply as Fe^{3+} ions in the clay structure. In the illite group, Fe^{3+} ions can replace Al^{3+} ions in the octahedral structure [11]. Fe_2O_3 is formed during sintering under oxidation conditions and from minerals in the clays, giving a reddish color to ceramic materials.

Feldspar: feldspars refer to a group of aluminum silicate minerals. The feldspar contained in the clays is a source of sodium and potassium oxides and plays an important role in ceramic materials with quality of flow agents, temperatures such as sintering temperatures, porosity after firing and facilitating phase formation [6]. The most representative are the orthoclase ($KAlSi_3O_8$) and albite ($NaAlSi_3O_8$).

Carbonates: calcium or magnesium carbonates can appear as coarse or small grains. If they are presented as large grains (>125 μm), they may not react completely and the resulting oxides may rehydrate causing expansion according to reactions [12, 13].

$$CaCO_3 \xrightarrow{\Delta} CaO + CO_2$$
$$CaO + H_2O \longrightarrow Ca(OH)_2 \tag{3}$$

3.3 Use of calcite in the ceramic and chemical industry

Ceramic enamels and frits: can be used in matte enamels as a source of CaO to form crystals such as wollastonite, anorthite, gehlenite or in transparent enamels giving shine.

Masses for ceramic coating: as a source of CaO up to the limit of 3%, $CaCO_3$ assists in the formation of the vitreous phase. CaO levels that vary from 8 to 14% favor the formation of crystalline phases such as gehlenite, wollastonite, pseudo wollastonite, and anortite.

Putties for limestone porcelain: calcium carbonates provide the CaO that are used as a flux in limestone porcelain masses.

Ceramic pigments: the calcium carbonate provides calcium oxide, which together with SnO_2 produces pink pigments.

Glasses: glasses based on NaOH and CaO use $CaCO_3$ in their composition.

Obtaining settlement mortars: as a plasticizing agent for water retention and aggregate incorporation.

Steel: $CaCO_3$ acts as a flux and pH regulator in water treatment and as lubricant for drawing steel rebars.

3.4 Specifications of raw materials for ceramic tiles

Sánchez et al. [14] defined some specification parameters for choosing raw materials for formulations of coating masses, as shown in **Table 2** below.

Calcium or magnesium carbonates can appear as coarse or small grains. If they are presented as large grains (>125 μm), they may not react completely, and the resulting oxides may rehydrate causing expansion.

In compositions of ceramic floor covering with low water absorption, $CaCO_3$ acts as a flux until the limit of 3%; above this value, $CaCO_3$ increases porosity and can be accepted up to 40% in porous coatings.

Product	(%) of carbonates	Max. particle size of $CaCO_3$ (μm)	Organic matter (%)	Sulfate content max. (%)	IP (%)
Stoned	≤3	≤125	≤0.3	0.2	20–40
Porous	≤40	≤125	≤0.3	0.2	20–40

IP: index of plasticity.

Table 2.
Specifications for choosing raw materials.

Enrique [15] recommends that the $CaCO_3$ particle size should be less than 125 μm, because particles of larger sizes, the CaO resulting from the dissociation of carbonates when calcined at 900°C, do not react with the SiO_2 present in the clays and feldspars that should form the pseudo-wollastonite and wollastonite phases, which can give rise to $Ca(OH)_2$ formed by the hydration of CaO, when the part comes into contact with the humidity of the air, generating problems of expansion by humidity, with consequent cracking.

The ceramic tile and brick industry have grown enormously in recent years in Brazil. The clays must have sufficient plasticity to provide mechanical resistance when forming by pressing, in order to guarantee the integrity of the piece in the path between the press and the oven. The feldspar contained in the clays are sources of sodium and potassium oxides, acting as fluxes at temperatures above 800°C for bricks and above 1100°C for ceramic tiles, which facilitates the formation of a vitreous phase and reduces porosity [16, 17].

Quartz is mixed with clay during geological formation. If it is present in a smaller proportion, it helps in the formation of the vitreous phase, in the degassing of organic matter and water. However, large proportions of quartz lead to a drastic reduction in mechanical strength after firing [18]. Iron oxide is present in ceramic raw materials in the form of hematite or goethite, giving the finished product a red color.

Calcite, which appears in most clays used in the production process of ceramic tiles of type BIIb, is a mineral that needs special care in its use due to its high loss to fire. When present in a proportion equal to or less than 3%, this mineral acts as a flux. However, in higher proportions, calcite can cause an increase in the final porosity of the product. In addition, the size of the calcite particle for processing ceramics must be less than 125 μm. For larger sizes, it is observed that the CaO resulting from the dissociation of carbonates can hydrate after burning, promoting variations in the dimension of the piece. Therefore, the use of limestone clays is a challenge, requiring care in processing and control in the formulation and burning of coatings. To ensure the correct sintering of the product, proper grinding and pressing of the raw material are necessary, in addition to efficient, fast burning with the lowest possible energy consumption.

4. Characterization of raw materials

4.1 Chemical analysis

Table 3 shows the chemical compositions of a typical Brazilian limestone clay used in ceramics [19]. The chemical compositions of the raw materials were determined by X-ray fluorescence spectroscopy by wavelength dispersion (WDFRX), in a Bruker S8 Tiger equipment, in which the percentages of constituent oxides were estimated by the method semi-quantitatively. For these measurements, samples with a mass of 10.0 g were pressed as discs with 40.0 mm diameter and 4.0 mm thickness. During measurements, the samples were kept in a vacuum of 10^{-6} bar. A mixture of P-10 (90% argon and 10% methane) was used in the proportional counter.

The results show that all clays are composed mainly of SiO_2 and Al_2O_3. These elements are associated with clay minerals, quartz, and feldspar structures [17]. The highest amount of SiO_2 was determined for sample C1. This component is important for the manufacture of ceramic tiles, as it improves workability and favors compaction. However, SiO_2 can also cause low mechanical strength of sintered ceramic bodies, in addition to reducing shrinkage during firing.

Oxide (%)	C1	C2	C3	C4
SiO_2	63.0	52.1	50.2	45.3
Al_2O_3	16.7	18.6	15.5	14.1
Fe_2O_3	4.7	6.8	6.2	7.1
CaO	0.9	2.1	7.2	12.7
K_2O	3.8	4.7	3.2	3.2
Na_2O	0.6	0.4	0.5	0.7
MgO	1.5	2.3	2.2	2.3
TiO_2	0.6	0.8	0.7	0.8
L.O.I	8.2	12.1	14.3	13.8

Source: Santos [19].

Table 3.
Chemical compositions of raw materials by X-ray fluorescence (XRF).

The amount of Fe_2O_3 detected in the samples was between 4.7 and 7.1%. These values are acceptable for use in ceramic tiles, such as bricks and tiles, this element being responsible for the reddish color of the sintered pieces as well as being a powerful flux [20]. The high content of calcium oxide in C4 (12%) and C3 (7%) stands out, characterizing these clays as limestone [21]. C4 clay was previously studied in Alcântara [16], which reports the formation of stains on the ceramic bodies produced with this material, after sintering at 1120°C. This behavior was associated with a high content of CaO, estimated at 10%, which during the burning phase, the dissociation of $CaCO_3$, promotes a high mass loss. C4 (13%) generates many pores, reducing water absorption and resistance of the final product. Thus, the higher the CaO content, the higher the $CaCO_3$ content and in addition, the higher the mass loss.

Analyzing the levels of alkaline oxides, it is observed that the sample C2 has the highest concentration of K_2O, while the concentration of Na_2O is approximately the same in the four samples studied. Alkaline and alkaline earth compounds have a melting effect, which facilitates the formation of liquid phase and linear shrinkage during burning [13].

Table 4 was arranged according to the increasing amount of CaO present in the clays. Note that C1 and C2 have CaO content below 3%. According to Enrique [15], CaO acts as a flux until the limit of 3% in masses of ceramic coating. The percentage of alkali oxides (Na_2O and K_2O), also presented in **Table 3**, is another major factor for the densification process, due to the great tendency of liquid phase formation during burning. Considering the sum of the percentages of CaO and alkali oxides in samples C3 and C2, it can be concluded that C2 has a higher proportion of fluxing oxides, suggesting that this sample is the most promising. On the other hand, clays

Clay	CaO (%)	$Na_2O + K_2O$ (%)
C1	0.9	4.4
C2	2.1	5.1
C3	7.2	3.7
C4	12.7	3.9

Source: Santos [19].

Table 4.
Percentage of fluxing oxides in clays determined via XRF measurement.

with a high limestone content, such as C3 and C4, tend to have greater porosity and less mechanical resistance after firing. Additionally, these two raw materials have lower alkaline oxide ratios than those observed for C3 and C2.

4.2 X-ray diffractometry

The X-ray diffraction patterns of the clays are shown in **Figure** 7 and correlate positively with the results observed by X-ray fluorescence. The X-ray diffractometry (XRD) technique was used to determine the crystalline phases. The samples were dried in an oven at 110 °C for 24 h, ground, and passed through a 150-μm mesh sieve. The diffraction patterns were obtained in a Rigaku D-MAX 100 equipment, using Cu Kα1 radiation (λ = 1.5418 Å). All measurements were carried out in the continuous scanning mode with speed of 1°/min, in the range of 5 to 65° and in the range of 2 to 15° in samples saturated with ethylene glycol for 1 h to identify montmorillonite by displacing the diffraction peaks at smaller angles compared to dry sample testing. The crystalline phases were identified through Match! (Phase Identification by Powder Diffraction) in the demo version, according to the ICSD (Inorganic Crystal Structure Database).

The main phases identified were quartz, kaolinite, muscovite, montmorillonite, calcite, feldspar, and hematite. Minerals from kaolinite and montmorillonite clay were identified in all analyzed clays. According to Celik [20], these clay minerals provide the necessary plasticity to guarantee conformation through the pressing process. The percentage of each crystalline phase present in the samples was estimated from the relative intensity of the main peaks in each phase. The values are shown in **Table 5**. The percentage of carbonates increases from 0.9% in C1 to 12.4% in C4.

4.3 Dilatometric tests

To verify the dimensional changes of expansion and thermal retraction of the samples, dilatometry tests were performed on a Netzsch dilatometer, model DIL

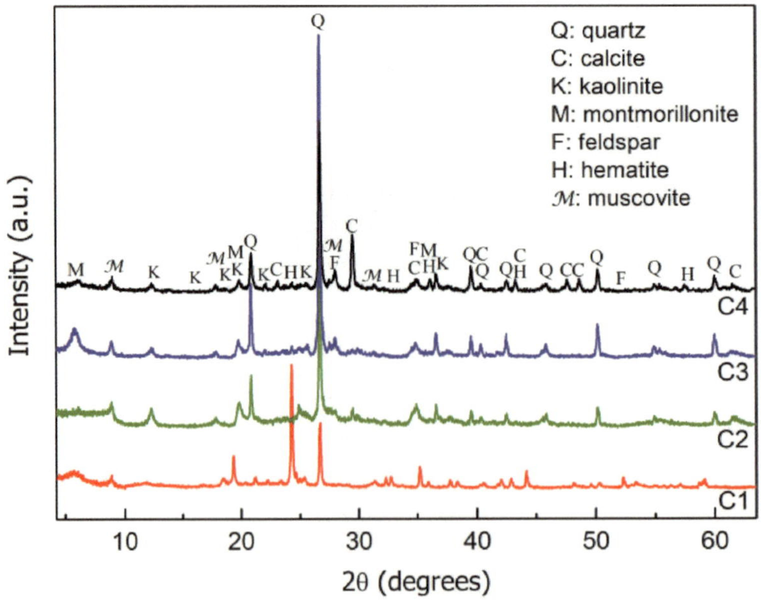

Figure 7.
X-ray diffraction patterns of the clays [19].

Minerals (%)	C1	C2	C3	C4
Quartz	55.7	51.8	65.1	57.1
Kaolinite	6.3	10.7	7.4	5.5
Muscovite	11.8	14.0	11.2	12.1
Montmorillonite	5.6	4.9	4.6	6.7
Calcite	8.6	2.8	1.1	13.7
Feldspar	6.3	9.9	6.2	3.2
Hematite	5.7	5.9	4.4	1.7

Table 5.
Mineralogical compositions of clays determined by XRD.

402PC, under synthetic air flow at 130 ml/min. For these analyses, the samples were compacted in a cylindrical shape, 12.0 mm in length and 6.0 mm in diameter. Under a constant heating rate of 10°C/min, the length of the compacted body is measured as a function of time and temperature, which varied from room temperature to 1150°C.

In **Figure 8** we can observe a slight expansion in all curves up to approximately 850°C, and at 573°C, the expansion was more pronounced due to the transformation of α quartz to β [22, 23], except for C2, which presents a lower percentage of free quartz. From 573°C, there was a gradual reduction in the expansion rate, occurring or starting with sintering, followed by an exponential retraction [22].

The results shown in **Table 5** with the percentages of CaO, Na_2O, and K_2O recommended by XRF measurements point out that sample C2 has a greater amount of funds (calcium carbonate up to a limit of 3% and alkaline oxides), or what is known as a greater linear shrinkage. Despite its advantages over the other samples, the C2 clay underwent deformation during firing up to 1150°C. This effect, known as pyroplastic deformation, may be due to the large proportion of funds in the sample, a high content of Fe_2O_3, and, even, the amount of organic matter [24]. One of the ways to control deformation during firing is to adjust the thermal cycle

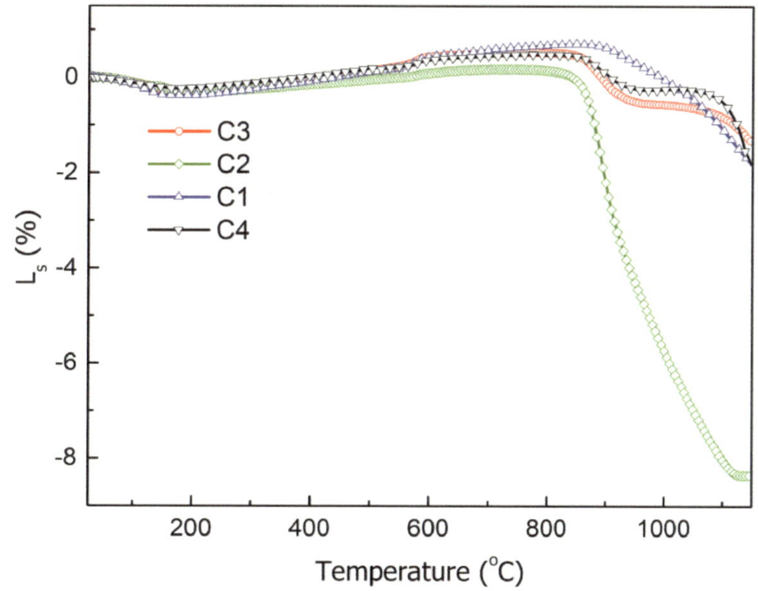

Figure 8.
Dilatometric curves of clays at a heating rate of 10°C/min [19].

through the dilatometric curves, so that the plate remains within the required standards [25].

4.4 Firing clays with limestone

Clays containing limestone when subjected to burning, $CaCO_3$ after heating, in the temperature range between 850 and 920°C, form CaO and release CO_2. An

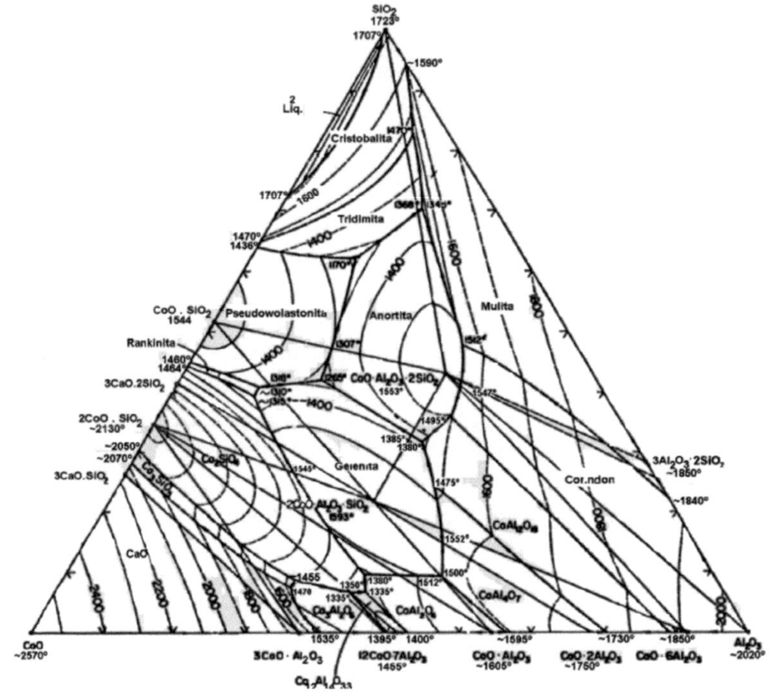

Figure 9.
Ternary diagram of CaO, SiO_2, and Al_2O_3.

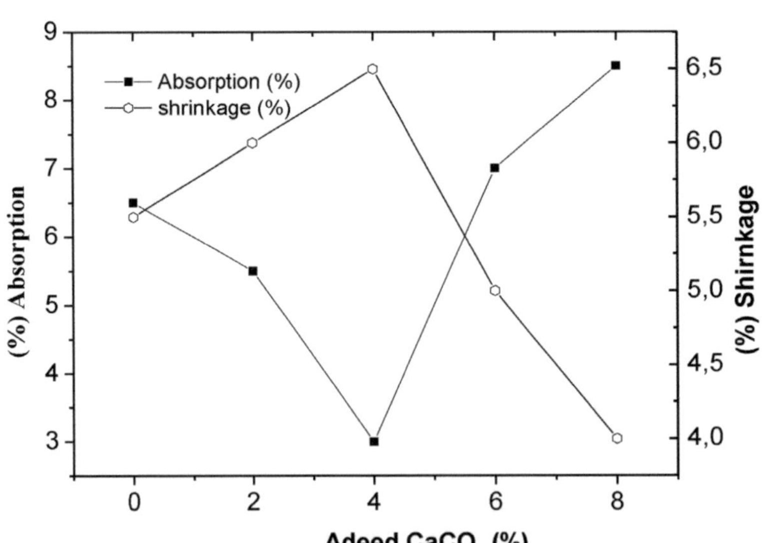

Figure 10.
Ceramic coating mass with incorporated calcite waste.

intense endothermic peak of approximately 35–44% of the mass loss can be observed in differential thermal analysis. In ternary diagrams, it is observed that there is a eutectic point (above 1170°C), which reduces the dimensional stability in ceramic products, which can melt quickly (**Figure 9**).

Clays when mixed with limestone can behave differently, as shown by Sánchez [25]. **Figure 10** shows a standard clay with 5 and 10% of incorporated limestone. It was observed that as the limestone and temperature increase, respectively, the dimensional instability increases. In other words, the retraction increases constantly, when it undergoes an exponential increase, reaching the melting point.

This phenomenon can be explained as follows: when exhibiting CaO up to the limit of 3%, this, associated with SiO_2 and Al_2O_3 present in clays and feldspars,

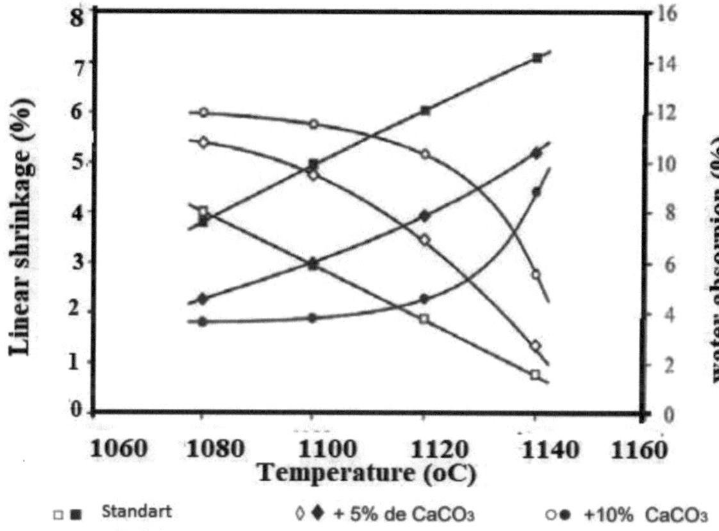

Figure 11.
Firing curve of a calcite clay.

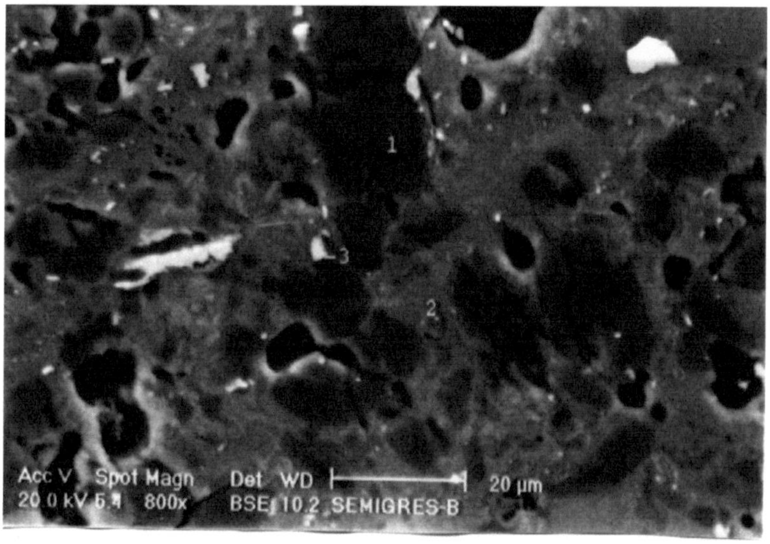

Figure 12.
Scanning electron microscopy of a ceramic with 10% of CaO.

helps in the formation of eutectic systems at 1170°C, with consequent formation of liquid phase and contributing to obtain the desired mechanical strength and porosity. When introduced in percentages above 4%, $CaCO_3$ levels are increased, and the composition moves from the eutectic line, forming crystalline phases such as $CaSiO_3$ (pseudo-wollastonite) and $2CaO{\cdot}Al_2O_3{\cdot}SiO_2$ (gehlenite). So, a larger number of pores is left by the eliminated CO_2. In this way, the porosity of the final product is increased, as shown in **Figure 11**. In **Figure 12** is shown a photo of a clay mass with 10% calibration in which the porosity exerted can be observed.

5. Conclusions and perspectives

Limestone is a contaminant for clay that above 125 μm can cause expansion and consequently cracks.

Rapid tests that mix clay with HCl can promote effervescence due to the release of CO_2 and contribute to decrease the amount of limestone.

In the ceramic industry, wet grinding of components is carried out in ball mills and grinding will be more efficient if the sieves are 150 to 325 μm. In ceramic mass formulations, the amount of CaO up to 3% contributes to the formation of the vitreous phase, however, between 8 and 14%, it favors the formation of crystalline phases, reducing the absorption of water and increasing the mechanical resistance.

Author details

Herbet Alves de Oliveira[1*] and Cochiran Pereira dos Santos[2]

1 Federal Institute of Sergipe, Estância, SE, Brazil

2 Physics Department, Federal University of Sergipe, São Cristóvão, SE, Brazil

*Address all correspondence to: herbetalves148@gmail.com

References

[1] Teixeira SR, Souza SA, Moura CAI. Mineralogical characterization of clays used in the structural ceramics industry in west of S. Paulo, Brazil. Cerâmica. 2009;**47**:204-207

[2] Vieira CMF, Monteiro SN, Dualibi FH. Consideraçoes sobre o uso da grunlometia como parâmetro de controle de uma argila sedimentar. Cerâmica Indusrial. 2005;**10**:23-26

[3] Gomes CF. Argilas o que são e para que servem. Lisboa, Portugal: Fundação Calouste Gulbenkian; 1988. (in Portuguese)

[4] Barba A et al. Materias Primas para la fabricación de suportes de baldosas cerámicas. 2nd ed. Castellón, Espanha: Instituto de Tecnologia cerâmica - ITC AICE; Castanèda; 2002. p. 292

[5] Padoa L. La Cottura dei Prodotti Ceramic. Vol. 65. Faenza: Faenza Editrice; 1979

[6] Castro RJS. Estudo do efeito do feldspato e resíduo de caulim na produção de revestimento cerâmico. Cerâmica Industrial. 2015;**20**(1):30

[7] Singer F, Singer S. Industrial Ceramics. USA: Chapmann Hall; 1995

[8] Santos PS. Tecnologia de Argilas. São Paulo: Editora Edgard Blücher; 1989

[9] Barba A et al. Materias primas para la fabricación de soportes de baldosas cerámicas. Espanha: Editora ITC instituto de tecnologia cerámica; 1997

[10] Mackenzie CR. The Differential Thermal Investigation of Clays. Londres: Editorial Cambridge; 1959

[11] Stepkowska E, Jefferis S. Influence of microstructure on firing color of clays. Applied Clay Science. 1992;**6**: 319-342

[12] Cargnin M, Souza SMAG, Souza AAU, Noni AJ. Determinação de parâmetros cinéticos da sinterização de revestimentos cerâmicos de monoqueima do tipo BIIa. Cerâmica. 2011;**57**:461-467

[13] López SYR, Rodriguez JS, Sueyoshi SS. Determination of the activation energy for densification of porcelain stoneware. Journal of Ceramic Processing Research. 2011;**12**:228

[14] Sánchez E, Ortz MJ, García-Tem J, Cantavella V. Efeito da Composição das Matérias-Primas Empregadas na Fabricação de Grês Porcelanato sobre as Fases Formadas Durante a Queima e as Propriedades do Produto Final. Cerâmica Industrial. 2001;**6**:15-22

[15] Enrique EJ et al. Decomposición de Carbonatos durante la Cocción de Piezas de Revestimento Cerámico Vidriado. Relacion com La Aparición de Pinchados. Qualicer; 1998

[16] Alcântara AC, Beltrão MS, Oliveira HA, Gimenez IF, Barreto LS. Characterization of ceramic tiles prepared from two clays from Sergipe-Brazil. Applied Clay Science. 2008;**39**

[17] Dondi M, Guarani I, Ligas GP, Palomba M, Raimondo M, Uras I. Chemical, mineralogical and ceramic properties of kaolinitic materials from the Tresnuraghes mining district (Western Sardinia, Italy). Applied Clay Science. 2010;**18**:145

[18] Seynou UM, Millogo Y, Ouedraogo J, Traor RK, Tirlocq J. Firing transformations and properties of tiles from a clay from Burkina Faso. Applied Clay Science. 2011;**51**:499

[19] Santos CP. Study of process variables and sinterization kinetics of materials used in the production of ceramic coatings [PhD thesis]. Brazil:

Federal University of Sergipe; 2016. Available from: https://ri.ufs.br/bitstream/riufs/3466/1/COCHIRAN_PEREIRA_SANTOS.pdf

[20] Celik H. Technological characterization and industrial application of two Turkish clays for the ceramic industry. Applied Clay Science. 2010;**50**:245

[21] Gonzalez F, Romero V, Garcia G, Gonzalez M. Firing transformations of mixtures of clays containing illite, kaolinite and calcium carbonate used by ornamental tile industries. Applied Clay Science. 1990;**5**:361

[22] Salem A, Jazayeri SH, Rastelli E, Timellini G. Dilatometric study of shrinkage during sintering process for porcelain stoneware body in presence of nepheline syenite. Journal of Materials Processing Technology. 2009;**209**:1240

[23] Zaied FH, Abidi R, Slim-Shimi N, Somarin AK. Potentiality of clay raw materials from Gram area (Northern Tunisia) in the ceramic industry. Applied Clay Science. 2015;**1**:112-113

[24] Amorós JL. Manual para el control de la calidad de materias primas arcillosas. 1ª Edición ed. Castellón, Espanha: Instituto de Tecnología Cerámica - AICE; 1998

[25] Prado ACA, Zanardo A, Moreno MMT, Menegazzo APM. Reduced susceptibility to pyroplastic deformation of clay at the Santa Gertrudes ceramic pole through the addition of raw materials. Cerâmica. 2008;**54**:7-11

Chapter 6

Multifunctional Clay in Pharmaceuticals

Nandakumar Selvasudha, Unnikrishnan-Meenakshi Dhanalekshmi, Sekar Krishnaraj, Yogeeswarakannan Harish Sundar, Nagarajan Sri Durga Devi and Irisappan Sarathchandiran

Abstract

Clay has its widespread applications in pharmaceuticals from ancient world to modern era. It is one of the excellent excipients present in the commercially available pharmaceuticals. Its use in many of dosage forms viz. in suspension, emulsion, ointments, gels, tablet and as drug delivery carrier as suspending agent, emulsifying agent, stiffening agent, binder, diluent, opacifier, and as release retardant have been explored in many studies. Variety of minerals is used as both excipient and as an active ingredient; among that kaolinite, talc, and gypsum are important. Their inertness, low toxicity, versatile physiochemical properties and cost effectiveness has increased its usage in pharma industries. Many minerals have its own pharmacological action as antacid, anti-bacterial, anti-emetic, anti- diarrheal agent and as skin protectant etc. Their unique structure which helps them to absorb material onto their layered sheets has opened a wide variety of applications in drug delivery. The understanding of surface chemistry and particle size distribution of clay minerals has led the pharmaceutical field in many directions and future perspectives.

Keywords: pharmaceutics, active pharmaceutical ingredient, excipients, inherent medicinal properties, drug delivery carrier, synergistic effect

1. Introduction

Usage of clay in medicine dates back to prehistorian era. Their evidences are present throughout the history from the clay pots of Nippur, Mesopotamia which gave the evidence of using clay against hemorrhages, the book "papyrus ebers" dating back in 1600 BC provided the details of using clay-based medicine for certain diseases. Many vital information on medical clay is mentioned in "On Airs, Waters and Places" written by Hippocrates (460–377 BC). One of the notable healing clay used in medicine during early days is known as Armenian bole (*bolus armenus*). They are pharmacologically used for the treatment of diarrhea, dysentery, hemorrhage, even as an astringent in few cases. Avicena in his book "El Canon" classified various types of clay and their internal and external applications. He also mentioned their role in anti-poison treatment and rheumatic disorders. Even though their consumption has been subjected to lot of questions the practice of using medicinal and edible clay prevails till date for their curative and beneficial effects.

Their usage in pharmaceutical industry is invariable due to their versatile property; they are used in almost every formulation like tropical to oral and also as an excipient. As the scientific community grows, there have been many papers published to support the medicinal and curative benefits of clay minerals [1–8].

The important component of clay is the clay minerals but it also comprises of associated minerals, organic, and inorganic materials. Clay can be grouped based on the geological aspects such as

- Primary or residual clay
- Secondary or sedimentary clay
- Special clay
- Common clay
- Refractory clay
- Nano clay
- Modified clay

Each clay has their own properties to distinguish themselves from the other. The classification also extends based on their geometrical shape, arrangement, and their usage. The classifications are given below.

Based on the geometry of the clay, it has been classified into four major groups and the subgroups are specified in **Table 1**.

1. Kaolinite
2. Smectite
3. Illite
4. Chlorite

The variation is due to the arrangement of tetrahedral and octahedral sheets, where kaolinite group has one tetrahedral sheet arranged over one octahedral sheet

S. No	General formula	Group	Layer type
1	$Al_2Si_2O_5(OH)$	Kaolinite-Serpentine	1:1
2	$Al_2Si_4O_{10}(OH)_2Mg_3Si_4O_{10}(OH)_2$	Pyrophyllitetalc	2:1
3	Montmorillonate $(Al_{1.67}Mg_{0.33})$ $Si_4O_{10}(OH)_{2M+0.33}$ Saponite:$Mg_3(Si_{3.67}Al_{0.33})O_{10}(OH)_{2M+0.33}$ Hectorite$(MgLi)_3(SiAl)_4O_{10}(OH)_{2M+}$	Smectite	2:1
4	$(Mg,Fe,Al)_3(Al,Si)_4O_{10}(OH)_{2.4}H_2O$	Vermiculite	2:1
5	$KAl_2(Si_3Al)O_{10}(OH)_2$	Mica/Illite	2:1
6	$Al_4[Si_8O_{20}](OH)_4Al_4(OH)_{12}$	Chlorite	2:1:1
7	$(Mg,AL,Fe^{3+})_5(Si,Al)_8O_{20}(OH)_2(OH_2)_{4.4}H_2O$ $Mg_8Si_{12}O_{30}(OH)_4(OH_2)_{4.8}H_2O$	Palygorskite-sepiolite group	

Table 1.
Subgroups of the clay.

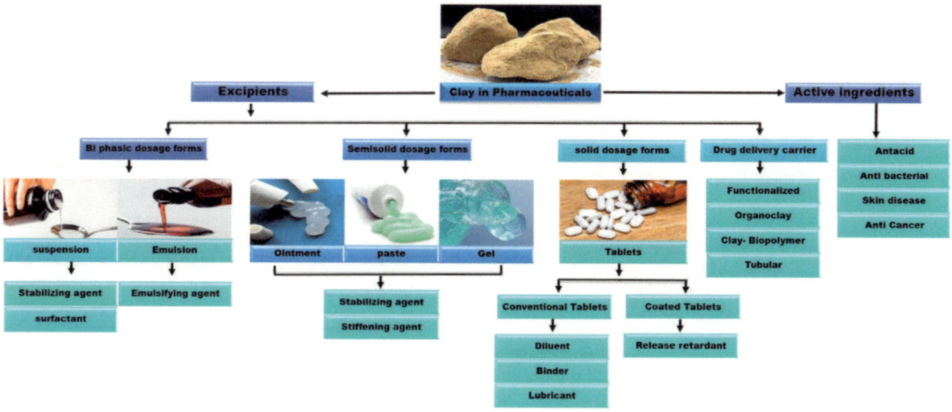

Figure 1.
Classification and usage of clay minerals.

whereas two tetrahedral sheets arranged over one octahedral sheet in smectite group. In case of chlorite group, octahedral sheet is arranged adjacent to 2:1 layer. The clay minerals exhibit versatile properties such as high adsorption capacity, chemical inertness, thixotropy, specific surface area, ion exchange capacity, less toxicity for oral administration, swelling property justifying their wide applications in pharmaceutical industries as excipient to enhance the physiochemical and organoleptic characteristics of a drug. It also helps in aiding drug conservation, elaboration, liberation of drug into the organism. These are achieved by incorporating the clay minerals such as disintegrates, lubricants, opacifiers, binders, diluents, isotonic agent, anti-caking agents, emulsifying agent, desiccant, thickening agent, and flavor modulators. Apart from the pharmaceutical applications, clay minerals also possess a lot of pharmacological properties like anti-bacterial, anti-viral, anti-diarrheal, gastro-intestinal protector, skin protection, and a potent detoxifier. The future trend holds for the MDDS using nano-clay minerals due to their inertness and biocompatibility. More details about the pharmaceutical and pharmacological uses of clay minerals and their advancement in drug delivery system have been discussed in the following sections. Further the classification and usage of clay minerals is illustrated in **Figure 1**.

2. Use of clay as excipients

An excipient is an inert additive ingredient formulated along with active compound to enhance its organoleptic, physiochemical properties. Clay has been used for almost every type of excipients (**Table 2**). Though several investigations showed that clay interacts with the drug molecule conversely to the nature of excipients, the interaction may hinder the drug bioavailability inside the body. The co-administration of montmorillonite leads to the degradation of certain cardiovascular tonic [9], anti-inflammatory drug [10]. Similarly, palygorskite and sepiolite degraded hydrocortisone and dexamethasone [11, 12]. Certain clay also affects the chemical stability of diazepam. Drugs such as phenobarbital sodium, diazepam solution, and lansoprazole show interaction with magnesite. Bioavailabity of tetracycline, indomethacin, aspirin, aspartame, ampicillin, cephalexin, and erythromycin has been drastically affected by calcium rich minerals. Clay minerals has also shown tendency to affect the drug liberation by interacting with the drugs through various mechanisms. They have shown to affect the liberation of amphetamines, analgesics, antibiotics, anxiolytics, solar protectors, and anti-histamines [13–16]. The adsorption of anti-histamines, antibiotics, atropine

Type of Excipient	Clay minerals	Drug used
As Diluent	Kaolin, talc, sepiolite, smectite, Magnesite	Slim well tablet, Quantrim, Riclasip, Riboflavin hard gelatine capsule
As Binder	Gypsum, hydroxyapatite, kaolin	Kaolin-Eudragit 30 in many tablets
As Disintegrant	Palygorskite, kaolin, sepiolite	Hydrochlorothiazide
As Granulating agent	Kaolin, anhydrite, periclase	Granules of NaCl and kaolin in Tablets preparation
As Emulsifying and wetting agent	Kaolin, palygorskite, smectites, sylvite, halite	Sulfur ointments.
As Suspending agent	Kaolin, palygorskite, smectites, sylvite, halite	Toxiban, morphine suspensions
As Drug delivery carrier	Natural MMT Modified MMT Acid treated MMT MMT-Saponite Kaolinite pillared MMT Functionalized kaolinite Natural MMT-Biopolymer MMT-Chitosan MMT-Alginate MMT-Guar gum MMT-PLGA MMT-PLLA MMT-PLLA-PEG-PLLA Modified MMT-Biopolymer Organomodified MMT-PVP/PCL MMT-Saponite-Chitosan Kaolinite-PVP-Sodium laural sulphate Natural HNT Natural Palygorskite Natural sepiolite HNT-Biopolymer HNT-Chitosan HNT-Alginate HNT-PEG HNT-PVA Sepiolite-Chitosan	Diclofenac sodium, chlorhexidine, gallic acid, promethazine Ciprofloxacin, Ag-Nanoparticles of Tetracycline, gentamycin, theophylline, nitric oxide, acetyl salicylic acid, Ibuprofen Mesalazine, oxytetracycline Venlafaxine, olanzapine Ibuprofen Dexamethasone, atenolol 6-mercaptopurin Gemcitabine Naproxen, curcumin Nicotine 5-flurouracil Chlorhexidine 5-amino salicylic acid, binase, tetracycline, amoxicillin, paclifaxel Ofloxacin, isoniazid Carvacrol, praziquantel Doxorubicin, aspirin, curcumin, Vancomycin, 5-flurouracil, hydrocortisone Atorvastatin

Table 2.
Application of clay minerals as excipient.

sulfate, salicylate, hyoscyamine, hydrobromide, paracetamol, and chloroquine into periclase and brucite has been showed by Khalil et al. [15]. These interactions have been proven useful since they are used to retard drug release. So they aid in controlled drug release and improve the Tmax considerably.

2.1 Clay in biphasic liquid formulation

2.1.1 In suspension

Excipients are required in the biphasic system in order to obtain proper wetting and to maintain stability of the formulation. In order to overcome the hydrophobicity of the drug and to aid in dispersion, clay minerals are added to the suspension as wetting agent. Sulfur ointment is prepared by blending kaolin with sulfur dispersed in oil phase [17]. The usage of bentonite as a wetting agent in foundation creams is also documented [18]. The clay also helps to maintain the stability by acting as suspending and anti-caking agent. They prevent sedimentation, changes in dispersion property, and flocculation. The criteria of selection of suitable suspending agent depend on compatibility, appearance, source, cost, and pH tolerance. The properties of the suspending agent including high viscosity at low shear rate, temperature, and storage tolerance, should not be affected by electrolyte or pH and be non-toxic. The formation of aggregates which in turn leads to the caking of solids would be the problem of suspension at the higher concentration. Reduction in the particle size or viscosity could not prevent caking. Caking can be prevented by flocculation and electrostatic stabilization [19, 20]. Kibbe showed that the increase in the stability of suspension using kaolin and talc as suspending and anti-caking agent. A suspension of pectin containing MAS dispersed in water along with kaolin under constant agitation at 70°C to which pectin was added, CMC was used to modify viscosity [17]. The usage of MAS and its four types (IA, IB, IC, and IIA) as a suspending agent has been commercialized and recognized by pharmacopeias, as they do not affect the pourability or spreadibility of the suspension. Sarfaraz [17] reported the usage of magnabrite S (10 mg/ml) and magnabrite K (15 mg/ml) in bismuth sub-salicylate suspension in which smooth gel was obtained as a final product. The usage of MAS (veegum HV) as gelling agent was studied by Sarfaraz [17] and Vanderbilt report [21]. Vanderbilt report also suggested that the gelling property of veegum HV was affected by acids and improved by alkalies. An antacid suspension with veegum HV was prepared by Sarfaraz [17] using xanthan gum to modify viscosity. Schott [22] optimize the concentration of bentonite as suspending agent and concluded that the concentration between 0.5–5% w/v was suitable for formulations. Bismuth subnitrate suspension produces good flocculation at 1.7% w/v of bentonite.

Many semisolid formulations use phyllosilicate as suspending agent due to their good adsorptive capacity which can be further improved by heating [23]. In many semisolid topical formulations, surface activated kaolin is added to enhance the stability and water miscibility of hydrophobic drug. Pharmaceutical preparation such as kaolin and morphine oral suspension BP, Toxiban suspension use kaolin as suspending anti-caking agent. The effect of crystallinity of kaolin on aqueous suspensions was studied by Ndlovu et al. Due to their dominant negative charge and ability to create permanent electrostatic repulsion justify kaolin use in suspension. The effect of non-ionic surfactant noigen RN10 (polyethylene alkyl phenyl ether) on kaolin wettability and stability was also studied. Clay can also help in stabilizing the suspension by having the effect on its rheological property since viscosity determines the rate of sedimentation according to stokes law. The different types and amount of clay are used to determine the final rheological property of the suspension. Dispersions showing dilant behavior contains 1:1 clay minerals and the pseudoplastic behavior is exhibited by 2:1 clay minerals. The commercialized MAS is a fibrous 2:1 clay containing blends of montmorellonite and seponite [6, 24, 25]. A combination of polyethylene glycol with hectorite improved the suspension stability [26, 27]. The modified hectorote such as quaternary C18 hectorite, Steralkonium hectorite is used in organic media to control viscosity [28].

The use of clay along with polymers has shown beneficial effect on their rheological properties, which has been demonstrated in griseofulvin suspension with MAS and sodium alginate by Dechow et al. [29] in sulfamethoxazole/trimethoprim suspension. The synergistic effect of CMC on properties of MAS such as viscosity, electrolyte tolerance, smooth flow property has led to the development of Veegum PLUS [30]. A similar synergism has also found in xanthan gum [31, 32].

2.1.2 In emulsion

Clay minerals are also added to pharmaceutical emulsion to prevent coalescence, creaming, phase inversion, breaking, and flocculation. This is due to their ability to wetted by the two liquid phases and be present as a barrier to prevent phase separation. Flocculation can be prevented by clay material due to their zeta potential. The stability of the emulsion is improved with increase in contact angle. They also used in the mechanical production of emulsion by acting as a surface acting agent which can bind in the interfacial layer but do not reduce the interfacial tension and interface. Due to its higher surface area, talc has been used as a emulgent in the cosmetic preparation [18]. The use of bentonite as emulsifying agent is familiar throughout the cosmetic industry. A nail enamel cream contains bentonite as an emulsifier was prepared by Carter [33]. He also proposed specific method of the cream preparation. Vanishing creams and skin protectants also use bentonite in a concentration of 2.5% w/v as emulgent [34]. In order to aid in easy application and adherent effect on the skin surface, clay minerals are added in corn and callus emulsion [34]. They are also added to hand creams as a thickening agent to retain better moisture control. The viscosity of the liquid eye liner formulation is maintained by addition of veegum [34]. Purified bentonite (Polargel NF) has been used as an emulsifying agent in cleansing lotion with HPMC and benzoyl peroxide. This has also been prepared as a cream by the addition of viscosity building agent (Carbomer) [17]. An anti-acne cream emulsion is prepared by using MAS [17]. MAS is also incorporated in cream emulsions for burns, methyl salicylate (analgesic effect), astringent zinc oxide, zirconium oxide and in zinc undecylenate lotions as a emulsifying agent [17]. Certain vitamins enriched skin creams also use MAS as emulsifier [30]. The synergic effect of xanthan gum along with MAS has been seen in zirconium oxide lotion [17]. Wenninger et al. [35] showed the usage of palygorskite as an emulgent (2–5% w/v). The presence of Na^+ and K^+ ions in halite and slyvite and their ability to control micelle size enumerate their use as emulsifying, thickening and anti-caking agent [36]. The long-term stability of pickering emulsions (where dodecane is used as oil phase) is improved by addition of kaolin (15% w/v) without any other additives [37].

2.2 Use of clay as an excipient in solid dosage forms

2.2.1 As diluent

Pharmaceutical oral preparations contain excipients such as diluent, flavorant, binder, disintegrant, pelletizing agent, granulating agent, sweetening agents, film coating agent, lubricant, and desiccants. Each excipient has its own influence on the formulation without interfering with active drug. It enhances the organoleptic and physiochemical property of the formulation. Diluents are selected based on the water solubility and bioavailability of active drug. Formulations with less water soluble drugs are incorporated with water soluble diluents and vice versa. Kaolin exhibit non-hygroscopic nature and low moisture content which determines its effectiveness as a diluent, as high moisture content may affect compressibility, physical and chemical stability of the formulation. Physiochemical parameters of kaolin have a direct impact over compressibility of the formulation [38]. The usage of kaolin as excipient for their adsorbent capacity has to be properly maintained since high adsorbent capacity can lead to less bioavailability of the drug [38–40]. Diluents are mainly added to the formulation to bulk up the volume and to facilitate easy compression. Diluents account for the 90% in low dose formulation. Product such as slim-well and quantrim use kaolin as bulking agent whereas mecysteine hydrochloride (Gastro resistant tablet) has heavy grade kaolin. Riclasip and co-amoxyclav DST grunenthal uses kaolin as adjuvant [41]. The use of kaolin with metronidazole (antibiotic and antifungal drug, Riazole) reduces its bioavailability, release characteristic and diffusion of drug inside the body [42]. The absorption of D, L-phenylalanine (analgesic and anti-depressant) onto the slurry of colloidal kaolin was showed by Bonner and Flores [43] through in vitro gross adsorption chromatographic study. Kaolin also affect the bioavailability of drugs like phenytoin (anti-seizure drug), promethazine-HCl (sedative and antiallergic drug), chloroquine (anti-malarial), propranolol (vasodilator), quinidine sulphate (cardiac antiarrhythmic drug), phenothiazines (trifluoperazine, fluphenazine, perphenazine, and thioridazine), guanethidine and hydralazine (antihypertensive drugs), procainamide and verapamil (antiarrhythmic drugs) with antidiarrheal Kaopectate® drug [44–49]. It has been reported that Langmuir isotherm was followed in drug absorption by kaopectate which extent the bioavailability but the rate of drug availability is retarded. The double layered adsorption pattern of mebeverine hydrochloride (antispasmodic drug) with kaolin and added electrolytes again follows Langmuir isotherm was studied by Al Gohary. These types of interactions of kaolin can be prevented by increasing the ionic strength of the drug solution and with the presence of $-NH_2$, –O–, and benzene ring, as chelating ligands led to the interaction. The presence of di-aromatic ring in naproxen (anti-inflammatory drug) and siloxane surface of kaolin are responsible for their interaction [50]. On other hand, the absorption of ampicillin and warfarin (anticoagulant) with antidiarrheal kaolin-pectin is shown to be unaffected by kaolin [51, 52], which was again conformed by Khalil et al. [53]. In fact, the use of kaolin as diluent in water soluble cationic riboflavin (vitamin B2) has improved the release rate of drug from hard gelatin capsule than any other diluents used. The rate of drug release is pH dependent [54]. The sustained release formulation of pyridoxine hydrochloride (vitamin B6) was prepared by using kaolin as diluent [55]. Vitamin drugs degrade easily in the presence of high moisture content since kaolin exhibit low moisture content the formulation containing vitamin B1 (thiamine) and Vitamin C (ascorbic acid) is more stable on addition of kaolin than other additives [56, 57].

2.2.2 As binders

Binders help to maintain the physical integrity of the solid dosage form owing to their mechanical strength. They also play a vital role in granulation, tableting, encapsulation by acting as a homogeneous dispersed matrix for adhesion of all material in the formulation. On this context, the mixture of kaolin-Eudragit (8% w/w) has been one of the good binders for tableting process. Eudragit is a poly ester based resin which exhibit hydrophilicity and does not get affected by varying pH and also the presence of kaolin help to obtain a uniform polymeric dispersion by differentiating water insoluble and hydrophilic particle within the system. Irrespective of their physiochemical properties, hydrophilic drugs are also blended with low water soluble drug in the kaolin-Eudragit formulation due to their larger permeability. Based on kaolin concentration they can also be used as film coating agent [58, 59]. Minerals like periclase, calcite, and magnesite are added as binders to increase the stomach pH [60].

2.2.3 As disintegrant

The release of drug from the formulation once it reaches inside the body mainly depends on the nature of the disintegrant used during formulation. Disintegrant facilitate breakdown of solid dosage form into smaller particulates. Poor solubility, poor gel forming capacity, good hydration capacity, good molding, flow property, and should not form complex are the required criteria for disintegrants. Both swelling property and ability to decompose at acidic environment of smectite made its use as a disintegrating agent. The presence of negative charge and ability to produce permanent negative surface charges helps kaolin as a disintegrant [61, 62]. A mixture of kaolin with surfactant and cellulose when added to formulation which already has starch as a disintegrant increased its shelf life over a long period of time [63]. Later, the use of kaolin as a positive effect than starch as a disintegrant has been proved [64].

2.2.4 As pelletizing agent

Kaolin proved its efficiency over bentonite by forming pellets which show faster disintegration while the pellets formed by bentonite was only erodible not disintegrable [65]. Kaolin along with biopolymer increase the drug dissolution rate of hydrochlorothiazide by forming pellets that rapidly disintegrated into the dissolution medium [66]. The primary aim of pelletizing agent is to form microspheres of uniform size which can be compressed into tablets or filled into capsules that rapidly disintegrates inside gastrointestinal drug where each pellet act as sustain formulation [67, 68]. Kaolin as a pelletizing agent in comparison with bentonite, talc, veegum and bentonite produce pellets with maximum yield, desirable size and smooth pellets on addition of SLS (5%) over the others [69]. The beneficial effect of crospovidone (5% w/w) and kaolin (25% w/w) with lactose as pelletizing agent for enhancing roundness and sphericity of pellets was demonstrated by Kristensen et al. [65]. The drug dissolution rate of riboflavin was higher with kaolin than microcrystalline cellulose and lactose [66]. Desirable size range and sphericity can be obtained by incorporating high kaolin content. Aerosil 200 (5%) along with kaolin (45%) has huge positive impact on sphericity of pellets [70]. The granulating agent is added to the formulation to improve flow property, density, appearance and uniform drug content, they also aid in compressibility of oral formulation. Wet and dry granulations are most commonly used method of granule preparation. Wet granulation involves wetting, nucleation, coalescence, breakage, and attrition process whereas dry granulation involves direct compression or slugging [71, 72]. Granules of desirable strength, size, cohesion, and uniformity can be produced by mixture of kaolin and sodium chloride (10% w/w) in wet granulation process. The comparative study of kaolin with polyethylene glycol and polyvinyl alcohol as a binder in calcium chloride showed kaolin PVA mixture gave larger yield and size than PGA kaolin mixture [73].

2.2.5 Aid in solubility, dissolution, and lubrication

Kaolin also helps to convert drugs form their crystalline to amorphous state to improve their solubility, dissolution rate, and bioavailability [74–76]. Kaolin was added to ibuprofen in order to convert it to amorphous salt from its crystalline form to facilitate higher dissolution and bioavailability in comparison with standard. Halite can be used to control osmolarity of the solution due to their high solubility in water [77]. Amorphization is inversely proportional to the kaolin concentration [74]. Clays are also used as desiccants due to their hygroscopic nature. Talc is used as lubricant and to prevent adhesion of powder to the compression pistons due to their soft and unctuous nature [78]. They are also used as flavorant to mask the taste of the formulation.

2.2.6 As coating agent

The use of film coating additive enhance the organoleptic characters of solid dosage form, helps in maintaining stability and control drug release profile [79, 80]. The decrease in the rate of drug release of diphenhydramine chloride, theophylline, and pseudoephedrine hydrochloride pellets coated with Eudragit on addition of kaolin (3:1 of resin) was studied by Ghebre-Sellassie et al. [58]. Kaolin is also added to the film coating of hypericon and kollicoat IR. Kaolin incorporated on the outer shell of triple pressed dyphylline coated tablets showed control release [81].

2.2.7 Enhancer of organoleptic properties

The organoleptic property of a drug can be modified on addition of excipients like pigments and opacifiers into tablets, capsules, syrups, and topical creams. These are necessary to avoid confusion while administering multiple medications and for easy identification of different dosages and help to protect the drugs from photo-oxidative damage. Clay minerals (calcite, rutile, hematite, and magnesite) possess a wide range of color from red, green, black, yellow, and white. Coloring E171 is most used pigment which is a synthetic analogue of white zincite. Synthetic rutile is used on sunscreen lotion as opacifier.

3. Use of clay as an active ingredient

Clay minerals also has its application as active ingredient in pharmaceutical preparation due to their ability to act as antacids, antianemics, mineral supplements, gastric protectors, laxative, antidiarrhoeaics, antibiotics, antiviral agents, wound dressing agent, detoxifier, antitumor agent, anti-inflammatory, and tropical analgesic.

3.1 Antacid

Acidity is caused by excess secretion of HCl in the stomach due to various conditions. Clay minerals overcome acidity either by neutralizing hydrochloric acid or by decomposition of minerals by absorbing H^+ ion on to their surface. Thus, restoring the stomach pH to 7 from 1.5 to 2.5. An effective antacid must increase the pH by three to four units and decrease the free acidity, which is seen in clay minerals like calcite, magnesite, periclase, brucite, and hydrotalcite. Whereas palygorskite, sepioite, montmorillonite, and saponite neutralize acidity by absorbing H^+ ion onto their surface. Their usage also can lead to certain side effects such as renal silica calculi, constipation in case of over accumulation of Ca^{2+} since they form insoluble hydrate phosphate and Mg^{2+} ion produces laxative effect but these effects can be avoided by using different mineral compositions. This combination has also an advantage of sustaining the drug release for example the co-administration of gibbsite with brusite prolonged its antacid action since brusite is fast acting and gibbsite is slow acting antacid [82].

3.2 Wound dressing agent

Wounds characterized by skin abrasion and vascular damage can lead to microbial invasion, toxicity, and even hemorrhagic shock due to uncontrolled bleeding.

This is countered by our body homeostatic response through coagulation of blood which prevents bleeding. Hemostasis follows sequential steps like: (1) thrombin formation is initiated, (2) activation and platelet aggregation (amplification), and (3) fibrin formation to stabilize the platelet clot (propagation). Hemostatic agents provide a physical mesh-like layer which aid in amplification and propagation steps of hemostasis thus leading to platelet aggregation and coagulation. The negative surface charge of kaolin at blood and plasma pH has a drastic effect on its blood clotting potential. The activation of blood coagulation factor XII to its active is done by kaolin on contact with blood and plasma. The active form of factor XII in turn activates factor XI and pre-kallikrein which helps in preventing bleeding. Therefore, many wound dressing products contain kaolin as topical hemostatic agents (Quickclot combat GauzeXL, Quickclotinterventional™) [83–87].

3.3 Peptic ulcer

Peptic ulcer is characterized by thinning of mucosal layer of the stomach due to the mucolytic activity of stomach enzymes, in order to reduce gastric irritation and provide a barrier for mucosal layer several clay minerals are used for their high sorption capacity and non-toxicity. These clay minerals absorb all the gases, toxins, bacteria and even viruses and reduce gastric secretions. They also act as protectants decreasing the glycoprotein degradation in the stomach. But their non-specific action has led to their minimal usage. Even though smectites prevent the pepsin damaging activity over the mucosal layer, their very less time of action and tendency to get degraded in the acidic medium has been a disadvantage but kaolin can be stable and show very low dissolution even at very less pH. They are taken as tablets, suspensions, or powders orally. They dissolve easily in the acidic medium aiding in their easy elimination and absorption [82, 88–90]. Kaolin due to their higher sorptive capacity delay gastric emptying and intestinal transit by enhancing triacylglycerol hydrolysis and promoting the intestinal uptake of non-esterified fatty acid and glucose [91].

3.4 Anti-diarrheal agent and anti-emetics

Kaolin has been used as an API in formulations for gastrointestinal like ASDA stomach upset tablets, Entroclam, or Boots kaolin [92, 93]. Diarrhea is caused by various factors like allergy, bacterial infection, intoxication, and low efficiency of intestinal sorption. It is characterized by increase in fluidity and frequency of evacuation. Anti-diarrhoeaics agents must have very good absorption capacity of excess water as well as gases in the digestive tracts. Activated clay minerals like kaolinite, palygorskite, sepiolite, and montmorillonite can be used against diarrhea for their high sorption capacity. They also prevent diarrhea by forming in soluble salts through release of Ca^{2+} (calcite) and Al^{3+} (gibbsite) ions [94–97]. Pharmaceutical products such as kaolin/pectin (Kaopectate®) and Kaomix® suspensions, kaolin Antacil®, Sainsbury's Diarrhea relief® and Treda® tablets contain kaolin as active ingredient against diarrhea due to hydrophilicity, surface area, microporosity, water osmotic and retention property as well as its antibacterial and antiviral effect (e.g., Norwalk and rotavirus, salmonella, Shigella and *Escherichia coli* bacteria) [98–101]. Minerals rich in Mg^{2+} or Na^{+} ion (mirabilite, epsomite, brucite, periclase, and magnesite) can act as laxative by increasing the osmatic pressure of the intestinal content which induces water level increase in the intestine and finally producing liquid feces. They are given as solutions, granules, and suspensions, and these ions are mostly excreted through fecal matter and a small amount through kidney or bile duct. Halite and sylvite are administered as saline through oral or parental route for the purpose of electrolyte replenishment (Na^+ and K^+). They are excreted through urine. Minerals rich in Cu^{2+} and Zn^{2+} (chalcocite, goslarite, and zincosite) can irritate gastric mucosa and trigger vomiting so they are used as direct emetic agents. When they reach intestine from stomach, they cause diarrhea. They are given orally as solutions. It can also be used to treat metal poisoning by removing them through vomit.

3.5 In anemia and inflammation

The use of clay minerals extents till its usage in disease like anemia. Anemia is caused due to less production red blood cells which may be due to less availability of Fe^{2+}. This can be treated with melanterite which is rich in Fe^{2+} ion and readily soluble in water. They are given as oral solution; these ions on reaching blood plasma convert into ferric ion by binding with a transferrin and globulin β. The excession is stored in liver, spleen, and bone marrow. Some are excreted through urine, gall bladder, and bile duct. Orally administered halite, epsomite, brucite, periclase, calcite, hydroxyapatite, magnesite, sylvite, melanterite as tablets provide ions such as Ca^{2+}, Na^+, Ca^{2+}, Fe^{2+}, K^+, PO_4^{3-}, and Mg^{2+} which are very much essential to our body [82]. Inflammatory response in our body is triggered to produce white blood cells and their mobility towards the injured site from infection through antigens or other harmful micro-organisms. Swelling, redness, pain, and heat are the main symptoms of inflammatory response. Lopez-Galindo and Viseras [102] presented the use of kaolinite poultices as anti-inflammatory drug, since they can absorb the excess fluid content near the infected tissue, which reduces pain and congestion considerably. They also aid in skin cooling by acting as a heat retention agent. Proper care of temperature must be taken while administering these dosages since it can have an effect over its therapeutic action [103–108].

3.6 Anti-bacterial

Minerals like sulfur, goslarite, borax, zincosite, chalcanthite, zincite, and alum are highly corrosive and toxic to pathogens in higher concentration hence they can be used as an antiseptic or disinfectant. They are also used as an astringent (chalcanthite), bacteriostatic agent (borax), fungicide, hemostatic agent (alum), and for skin damage. The bactericidal activity of clay extents to many drug resistant bacteria like *Pseudomonas aeruginosa, E. coli, and Staphylococcus aureus* due to their physical and chemical properties that help them to envelope bacterial cells and interrupting their nutrient uptake, this is due to their high surface attraction towards the bacterial cell wall. Ions present in clay minerals also play an important role in their bactericidal property. Divalent cations like Cu^{2+} and Fe^{2+} are easily transferred and oxidized inside the bacterial cell to produce intercellular hydroxyl radicle which are lethal to them. The tri- or tetravalent cations show their activity inhibiting the influx or efflux pumps. Many modified clays have been reported to have good bactericidal activity, for example the photocatalytic activity of zinc oxide and Ti make TiO_2 (ZnO)/kaolinite make the effective against *Enterococcus faecalis, E. coli, and Pseudomonas aeruginosa.* Moreover, *Pseudomonas aeruginosa* is also susceptible to kaolin modified with CTAB and Cu. Oral pathogens like *E. coli, Bacillus subtilis, and klebsiella pneumonia* are effectively killed by kaolin/iron-porphyrin hybrid. Nano-composite of silver-kaolinite also demonstrated to have antibacterial property [109]. They use clay as an adsorbate to remove pathogenic viruses and certain phages are under investigation since the twentieth century. First it is thought that viruses by electrostatic interaction are adsorbed onto the clay surface due to their valency associated cations and cation exchange capacity but later studies showed that amino acid and carboxylic residue present on the outer shell viral protein coat is responsible for the net charge and their adsorption into clay minerals depends on pH, ionic strength, and isoelectric points of the clay and virus. Further studies show that hydrophobic interactions of crystallized kaolin show greater affinity for certain bacteriophages. In vitro studies of kaolin against hepatitis C virus, human enteric pathogenic adenoviruses (hAdVs) and adenovirus (HAdV-5) proved kaolin to be an effective material that could be used against certain viral infections [110]. Many substances like heavy metals, toxins, mycotoxins, and overdosed drug compounds can be removed from the gastrointestinal tract by administering kaolin as a detoxifying agent.

3.7 Miscellaneous

Kaolin is found to influence superoxide radicle generation by immunocompetent carcinoma blood cells of Lewis lung in mice this opens a huge area of application for kaolin usage in restorative cancer treatment. Their antitumor property has made it a potential candidate under investigation [92, 100, 110–112]. Minerals like zincite, talc, rutile, hydrozincite, smithsonite, kaolinite, and smectites are used as an protective agent in skin to prevent against certain external environmental condition and pathogens. They have suitable properties like high sorption, non-cytotoxic, little antiseptic and bactericidal as discussed above. They are used as creams, powders, and ointments (REINOL drygard, DP1, kerodex 51, etc.). They are also incorporated in sunscreen formulation to prevent against harmful UV-A and UV-B. Minerals like rutile and zincite absorb, reflect, and scatter the radiation but they may cause skin damage by photocatalytic action this can be overcome by using kaolin as protectant since it shows higher UV protection capacity due to the high Fe_2O_3 content. Kaolin also used to absorb excess moisture, oily secretion, surface lipids, and superficial toxin from the skin surface to prevent acne, blackheads, bacterial, and fungal infections. Even they are used for insect bites to give relieving effect [109].

3.8 Skin protectant

Sulfur containing minerals are extensively used as keratolytic reducers as they are effective against dermatitis, eczemas, and psoriasis. Sulfur reacts with cysteine in the presence of keratinocyte producing hydrogen sulphide which breaks down keratin. Dandruff is treated with cadmium sulphide shampoo. The less common adverse effect of sulfur applied as topical cream is an added advantage [82]. Isotonic collyrium contains dissolved halite which is used as decongestive eye drops for treating eye dryness, irritation, and other ocular discomforts.

4. Use of clay as a drug delivery system

An effective drug delivery system is essential for achieving proper bioavailability of the drug administered. Recent studies have paved way for many new modified drug delivery system that has led to sustained drug deliver, controlled drug delivery, and site-specific drug delivery (**Table 2**). Each has its own mode of release and application in the body. The modification can be made only through excipients employed. The proposed excipients must possess good delivery efficiency at the same time have inertness, easy availability, cost effective, and low toxicity. Thus, all these are readily available in clay minerals and their physiochemical properties make them a potential candidate for design of MDDS. Clay minerals either in their native form or modified in certain way to improve their physiochemical properties to aid their use as a drug carrier for delivering system.

4.1 As release retardant

Kaolin (1:1) and smectite (1:2) are the most common used clay groups in the design of MDDS due to their geometrical structure. The side effects like short half-life and requirement of frequent dosage in diclofenac sodium (NSAID) can overcome by intercalating it with MMT that prolong drug release. The toxicity of topically administered chlorhexidine (antibiotic) can be prevented by using Na-MMT as a drug delivery carrier. The photolytic damage of promethazine (antihistamine) when administered topically is reduced by intercalation of drug with Na-MMT. The in vitro activity and controlled release of paclitaxel (anti-cancer) is increased by intercalation with Na-MMT and coating with chitosan (biopolymer). Gallic acid has various properties like poor solubility, permeability, and faster metabolism which make it difficult for dose administration and drug release. The idea of Gallic acid with MMT was suggested and carried to and characterized the drug release profile which showed promising results. A dermal patch prepared using MMT-Na loaded with silver (antimicrobial agent), lidocaine (mild analgesic), and betaine hydrochloride showed a controlled release of lidocaine. The controlled drug release of metformin with Na-MMT was studied in order to reduce the side effect and dosage of drug. But the study concluded that the drug release was highly pH dependent and needed further analysis. MMT enhanced the antibacterial activity with TiO_2 coated with alginate against both gram positive and negative bacteria. Mesalazine (5-Aminosalicylic acid) must reach colon to show its therapeutic action against Crohn's disease but mesalazine is highly absorbed in the acidic environment of stomach. This can be altered and slow drug dissolution in stomach was achieved using MMT-Na encapsulated into alginate beads. Acidification improves clay surface area and increases pores for sorption. These acidified clays was used as a carrier to deliver ciprofloxacin, the acid treatment retarded the drug release due to the changes in the interlayer charges, this also provided insight on the usage of interlayer charge modifications of the clays can be a useful phenomenon for drug delivery. Silver nanoparticles are also loaded with modified clay and their antibacterial activity is characterized and a comparative study between modified and unmodified clay hybrid for their loading capacity and antibacterial activity against *S. aureus* and *E. coli* was done. The results showed that both the loading capacity and antibacterial activity was higher for acidified clay when compared to its native form. Periodontal extended release of tetracycline was achieved by intercalation of drug with MMT hybrid (Veegum HV) and chitosan as mucoadhesive base, the formulation required only once per week. Optimization of gentamycin loaded with another clay hybrid (Veegum F) was done and its antibacterial activity was evaluated, the formulation a delayed drug release up until 8 days. Theophylline was loaded onto MMT hybrid (Veegum F) to prevent the premature gastric drug absorption and to give a sustained release in intestine. Electrostatic interaction with clay hybrid prevents theophylline absorption at the stomach pH and facilitates slower absorption on the intestinal pH. The in vitro antibacterial activity of ciprofloxacin was determined by the type of interaction it forms with the clay hybrid, a weak interaction ensures easier release of drug into the desired site. Another modification of clay minerals which is used in MDDS is functionalization of the interlayers. One such functionalized clay is pillared clay which have large specific surface area and larger porosity due to the cationic exchange with inorganic compounds (4-(dimethylamino)-1-(4-vinylbenzyl) pyridiniumchloride and 1-methyl-3-(4-vinylbenzyl) imidazolium chloride). Ibuprofen loaded into MMT hybrid pillarized with Fe^{3+} and Fe^{2+} showed a delay in drug release under various physiological conditions. Doxorubicin (anticancer) was loaded into functionalized kaolin showed good drug loading efficiency and therapeutic action.

4.2 Clay-biopolymer combination

Biopolymers can also influence drug delivery by encapsulation or by surface coating over the formulation and help in controlled drug delivery. Hence, clay hybrids are coated with biopolymers to enhance their action. Clay hybrid and biopolymer (chitosan) showed synergistic effect on loading efficiency and drug release of 5-aminosalicylic acid. The oral bioavailability of oxytetracycline (broad spectrum antibiotic) was improved by loading over chitosan-MMT carrier. The cytotoxicity of chlorohexine to fibroblast was reduced by preparing a film carrier made of MMT-chitosan complex. The prepared topical formulation showed a controlled release of drug and reduction of cytotoxicity was also reported. MMT-chitosan glutamate composites also reduced the cytotoxic effects of silver sulfadiazine (skin burns). The result show that electrostatic interaction of MMT with polymer helps improving the drug absorption and the increase an antibacterial activity of formulation was also noted. Apart from chitosan, other biopolymers have been used to prepare drug carriers. A combination of guar gum-MMT hybrid was used to prepare a controlled release formulation of ibuprofen to reduce their side effects on intestinal tracts. The need of frequent dosage of venlafaxine (anti-depressive drug) was reduced by preparing beads with crosslinking of sodium ALG with drug-MMT in $CaCl_2$. Olanzapine (schizophrenia and bipolar disorder) was incorporated into different polymer composition and their drug release at different pH was studied and the results were in comparison with the marked drug and an effective controlled release was obtained by cloisite-drug with a blend of polymers (Alginate and xanthan gum). The solubility of curcumin (anticancer, anti-inflammatory and antibacterial agent) can be increased by dispersing the drug with CMC and loading into MMT, the role of MMT here is to improve the drug release in the acidic environment. A transdermal DDS was prepared from MMT nanocomposite, pectin, and methyl cellulose which is used to load ketorolac (NSAID). The formulations showed immediate release of drug from the nanocomposite layer but the increase in MMT showed controlled drug release. The PLA microspheres of 6-mercaptopurine (anticancer drug) with MMT showed a faster release rate and increased the drug solubility. The presence MMT also helps to control the drug release. The performance of a pH dependent swelling polymer (poly acrylamide-co-maleic acid) was improved by the addition of MMT hybrid, it exerted a control over the caffeine drug release even during sudden pH transitions. A prolonged oral DDS was prepared for 1,3,4-oxa(thia)diazole (antifungal, antibiotic, analgesic and anti-inflammatory agent) by preparing nanocomposite using MMT hybrid. A site specific DDS was developed using MMT-polymer hybrid for anticancer treatment by co administration of doxorubicin and methotrexate with ciprofloxacin (antibiotic). Specificity of the DDS depends on the pH of the tumor cells. The results showed that the entire three drugs exhibit delayed drug release, where the anticancer drugs showed similar release profile and ciprofloxacin showed a different release profile at pH 5.8 and 4. PLGA is another biopolymer which is extensively used as carrier for drug administration. A double emulsion of atenolol with PLGA and MMT was prepared by Lal and Datta to increase the half-life and dissolution rate of the drug. The result suggested a controlled drug release in both acidic as well as basic medium with marking changes in the acidic medium. The hydrophobic drug (dexamethasone) was also intercalated with PLGA and MMT by Jain and Datta to lower the risk of side effects and achieve effective plasma concentration at minimal dosage. PLGS-MMT nanocomposite is used as a carrier for insulin. In vitro studies by Lal et al. suggested the protective nature of MMT even in acidic conditions and also they do not affect the HEK-293 cells growth. A triblock (PCLA-PEG-PCLA) copolymer hydrogel of MMT-gemcitabine (anticancer) was prepared for the intravenous administration of drug since the drug is metabolized rapidly and require high dose. In vitro drug release studies suggested that MMT significantly reduced the drug release and the side effects. Nanocomposite of MMT hybrid with HEMA was used to modify the dissolution profile of paracetamol by Bounabi et al. The inclusion of MMT improved the drug encapsulation of paracetamol in the PLA-drug nanocomposite. A site specific DDS was prepared for doxorubicin was prepared using MMT hybrid with PE-5000/PEG750 polymer. Organoclay (MMT-PVP hybrid) nanocomposite was used to encapsulate copaiba oleoresin a natural derivative used against endometriosis. The nanocomposite showed effective controlled drug release at acidic pH. A polymeric composite (PVA, CS and MMT) was prepared to encapsulate 5-flurouracil (anticancer) in order to compensate its poor oral absorption and rapid metabolism. The results also indicate that the drug loading efficiency and drug release depends on the MMT concentration. Clay minerals like halosite and fibrous phyllosilicates have MMDS application due to their tubular and ribbon shaped structure. For example, the Hal nanotubes can adapt to any morphology making them used for wide variety of applications in MDDS. Various antibiotic like ciprofloxacin, chlorpheniramine, tetracycline, diphenhydramine has been loaded on to the hal nanotubes and investigated. Cationic exchange capacity and pH determines the drug loading capacity on to the hal nanotubes. Thermodynamic equilibrium also affect the drug loading in hal nanotubes as in the case of isoniazid (antituberculosis drug). The immobilization of binase (RNase enzyme) which is used in the genetic treatment of cancer was done and an enhanced anticancer property was reported. Vancomycin and breviscapine has been loaded on to hal nanotubes by vacuum cycle and the resultant complex showed a sustained release of drugs. Amoxicillin loaded onto Hal nanotubes are combined with biopolymers (PLGA and Chitosan) and the drug release was studied the results suggest a sustained release is obtained on both formulations with and without biopolymers than biopolymer-drug complex. PMMA was coated onto paclitaxel-hal nanotube complex to improve the anticancer activity of the drug. Volatile drugs are also adsorbed onto the clay minerals which help in preventing the evaporation of those drugs and retaining their therapeutic action. The volatile drug absorption of MMT, hal nanotubes and palygorskite was evaluated by loading carvacrol (treat skin lesions). Good adsorption was seen in palygorskite. Veegum HS and sepiolite improved the solubility and dissolution rate of praziquantel (treatment of schistosomiasis) in both acidic and basic environment. Higher dissolution was achieved by oxaprozin (Non-steroidal anti-inflammatory agent) on mixing it with clay hybrid of palygorskite and sepiolite modified with cyclodextrin. Curcumin was loaded with both functionalized clay and cellulose-MMT complex to improve its site specific action and synergic effect on wound healing. A phospholipid nanocomposite of hal nanotubes was prepared to achieve a sustained release of doxorubicin. The electrostatic interaction and intermolecular hydrogen bonding between palygorskite and chitosan was studied for its usage as a drug carrier for 5-aminosalicylic acid. Sepiolite is also used with chitosan as drug carrier for tetracycline and cefazolin. In vitro studies show the swelling of gel is facilitated by chitosan whereas the drug release id controlled by crosslinking of polyvinylacryalate (PVA). The synergistic effect of hydroxypropylmethylcellouse acetate succinate along with atorvastatin and celecoxib for colon cancer is due to their effective solubility in basic pH. The controlled release of drug in colon is achieved by the preparing microspheres of hal nanotubes and HPMCAS loaded with anticancer drugs.

5. Pharmaceutical clay in share market

The clay minerals are exported worldwide for their various applications in construction and pharmaceutical preparations. Based of the application the clay mineral export is classified as tableware, sanitary ware, medicinal applications. The key minerals are bentonite and kaolin which accounts for major export. Bentonite is exported as sodium, calcium and sulfur bentonite. The global market demand of bentonite (**Figure 2a** and **b**) was 22.68 million metric ton by 2016 and estimated to be 25.15 Million metric ton by 2021 with a CAGR increase of 2.12%. The global market share of bentonite by 2017 was 1.43 billion and estimated to be increasing scale due to the market demand and increase in the applications of clay minerals. The major region of export is classified as Asia Pacific, North America, Europe and Rest of the world with Asia and North America accounting for most. The major companies exporting clay minerals are Ashapura groups (India's major exporter), Imerys(sandB), Taiko group, Huawei Bentonite, Theile kaolin company, Kaolin A.D, J.M.Huber, Daleco resources.

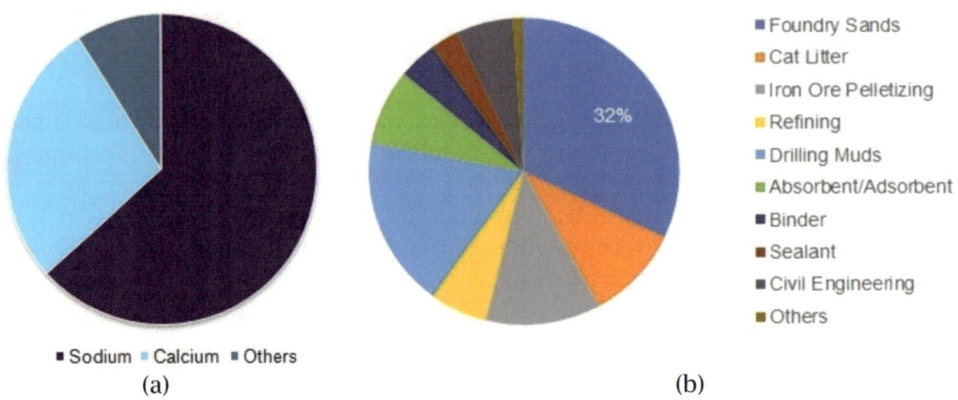

Figure 2.
(a) Bentonite classification based on minerals. (b) Bentonite usage on global demand scale.

6. Conclusion and outlook

From the moment of discovery clay minerals have been immensely useful to human life in both ceramic and health. Their usage in health care has made it a essential compound in many pharmaceutical preparations. Their inertness, low toxicity, versatile physiochemical properties and cost effectiveness has increased its usage in pharma industries. At the same time precautions must be taken while incorporating higher doses of clay and while co-administering clay with drug. Since some clay has been reported to reduce the efficacy and bioavailability of certain classes of drugs like antacid and in higher doses it might cause tissue toxicity. The understanding of surface chemistry and particle size distribution of clay minerals has led the pharmaceutical field in many directions and future perspectives. Their unique structure which helps them to absorb material onto their layered sheets has opened a wide variety of applications in drug delivery. Their ability to control and alter drug release profile can been exploited in many ways to design a effective drug delivery system. Further advancements in nanotechnology have helped to synthesize and modify this clay mineral to enhance their physiochemical properties and their usage as excipient. Though clay and their minerals are used in its natural state for drug delivery, some require additional modification for their usage and this modification plays a key role in determining the economical aspect of drug designing. The development of machinery which helps us to understand better about various unknown properties of clay minerals which were not understood before will aid us to utilize clay minerals in various other applications.

Author details

Nandakumar Selvasudha[1]*, Unnikrishnan-Meenakshi Dhanalekshmi[2]*, Sekar Krishnaraj[3], Yogeeswarakannan Harish Sundar[3], Nagarajan Sri Durga Devi[3] and Irisappan Sarathchandiran[1]

1 School of Pharmacy, Sri Balaji Vidyapeeth Deemed to be University, Puducherry, India

2 Oman Medical College, Bowshar Campus, Muscat, Sultanate of Oman

3 Department of Biotechnology, Anna University, Chennai, India

*Address all correspondence to: nkselvasudha@gmail.com
and dhanamnair2013@gmail.com

References

[1] Novelli G. Applicazionimedicali e igienichedellebentoniti. In: Veniale F, editor. AttiConvegno "Argille Curative", SaliceTerme/PV. Gruppo Italy. AIPEA; 1996. pp. 25-43

[2] Robertson RHS. Cadavers, choleras and clays. Mineralogical Society of Great Britain & Ireland. 1996;**113**:3-7

[3] Carretero MI. Clay minerals and their beneficial effects upon human health. A review. Applied Clay Science. 2002;**21**:155-163

[4] Veniale F, Barberis E, Carcangiu G, Morandi N, Setti M, Tamanini M, et al. Formulation of muds for pelotherapy: Effects of "maturation" by different mineral waters. Applied Clay Science. 2004;**25**:135-148

[5] Viseras C, Lopez-Galindo A. Pharmaceutical applications of some Spanish clays _sepiolite, palygorskite, bentonite: Some preformulation studies. Applied Clay Science. 1999;**14**:69-82

[6] Viseras C, Aguzzi C, Cerezo P, Lopez-Galindo A. Uses of clay minerals in semisolid health care and therapeutic products. Applied Clay Science. 2007;**36**:37-50

[7] Gomes CSF, Pereira Silva JB. Minerals and Human Health. Benefits and Risks. Centro de Investigação "MineraisIndustriais e Argilas". Fudação para a Ciência e a Tecnologia do Ministério da Ciência, Tecnologia e Ensino Superior. Aveiro (Portugal); 2006

[8] Carretero MI, Gomes C, Tateo F. Clays and human health. In: Bergaya F, Theng BKG, Lagaly G, editors. Handbook of Clay Science. Amsterdam: Elsevier; 2006. pp. 717-741

[9] Porubcan LS, Born GS, White JL, Hem SL. Interaction of digoxin and montmorillonite: Mechanism of adsorption and degradation. Journal of Pharmaceutical Sciences. 1979;**68**:358-361

[10] Forteza M, Galan E, Cornejo J. Interaction of dexamethasone and montmorillonite. Adsorption – degradation process. Applied Clay Science. 1989;**4**:437-448

[11] Cornejo J, Hermosin MC, White JL, Barnes JR, Hem SL. Role of ferric iron in the oxidation of hydrocortisone by sepiolite and palygorskite. Clays and Clay Minerals. 1983;**31**:109-112

[12] Forteza M, Cornejo J, Galan E. Effects of fibrous clay minerals on dexamethasone stability. In: Konta J, editor. Proceedings of the Tenth Conference on Clay Minerals and Petrol, Ostrava. Prague: Universitas Carolina; 1988. pp. 281-286

[13] McGinity JW, Lach JL. Sustained-release applications of montmorillonite interaction with amphetamine sulfate. Journal of Pharmaceutical Sciences. 1977;**66**:63-66

[14] Porubcan LS, Serna CJ, White JL, Hem SL. Mechanism of adsorption of clyndamicin and tetracycline by montmorillonite. Journal of Pharmaceutical Sciences. 1978;**67**:1081-1087

[15] Iwuagwu MA, Aloko KS. Adsorption of paracetamol and chloroquine phosphate by some antacids. The Journal of Pharmacy and Pharmacology. 1992;**44**:655-658

[16] Del Hoyo C, Vicente MA, Rives V. Application of phenyl salycilate–sepiolite systems as ultraviolet radiation filters. Clay Minerals. 1998;**33**:467-474

[17] Sarfaraz N, editor. Handbook of Pharmaceutical Manufacturing Formulations. Cambridge: CRC Press; 2004

[18] Gabriel DM. Vanishing and foundation creams. In: Harry RG, editor. Harry's Cosmeticology. The Principles and Practice of Modern Cosmetics. Vol. I. 6th ed. London: Leonard Hill Books; 1973. p. 83

[19] Mathews BA, Rhodes CT. Use of the Derjaguin, Landau, Verwey and Overbeek theory to interpreter pharmaceutical suspension stability. Journal of Pharmaceutical Sciences. 1970;**59**:521-525

[20] Russel WB, Saville DA, Schowalter WR. Colloidal Dispersions. Cambridge: Cambridge University Press; 1995. p. 525

[21] Vanderbilt Report. Technical Information: "VEEGUM-The Versatile Ingredient for Pharmaceutical Formulations". R.T. Vanderbilt Company Bulletin No. 91R. 1984. Available from: www.rtvanderbilt.com

[22] Schott H. Controlled flocculation of coarse suspensions by colloidal dispersed solids I: Interaction of bismuth subnitrate with bentonite. Journal of Pharmaceutical Sciences. 1976;**65**:855-861

[23] Sweetman SC. Martindale: The Complete Drug Reference. 34th ed. London: Pharmaceutical Press; 2005. p. 2756

[24] Viseras C, Meeten GH, López GA. Pharmaceutical grade phyllosilicate dispersions: The influence of shear history on floc structure. International Journal of Pharmaceutics. 1999;**182**:7-20

[25] Viseras C, Cerezo P, Meeten GH, López-Galindo A. One dimensional filtration of pharmaceutical grade phyllosilicate dispersions. International Journal of Pharmaceutics. 2001;**217**:201-213

[26] Omar SM, El-Nahhas SA, Khalil RM, Salama HA. Effect of polyethylene glycols on the rheological characteristics of Macaloid dispersions. Journal of Drug Research. 1994;**21**(1-2):91-103

[27] Ash M, Ash I. Handbook of Pharmaceutical Additives. 2nd ed. Endicott: Synapse Information Resources; 2002. p. 487

[28] Clarke MT. Rheological additives. In: Laba D, editor. Rheological Properties of Cosmetics and Toiletries. New York: Marcel Dekker; 1994. pp. 55-152

[29] Dechow HJ, Von Dölcher D, Hübner G, Kim S, Lämmerhirt K, Pich CH, et al. Zurentwicklung von oralesdareichungsformen der kombinationsulfamoxol/trimethoprim (CN3123). Arzneimittel-Forschung. 1976;**26**:596-613

[30] Vanderbilt. Technical Information. 2006. Available from: www.rtvanderbilt.com

[31] Kovacs P. Useful incompatibility of xanthan gum with galactomannans. Food Technology. 1973;**27**(3):26-30

[32] Ciullo PA. Rheological properties of magnesium aluminum silicate/xanthan gum dispersions. Cosmetic Chemist. 1981;**32**:275-285

[33] Carter HM. Fingernail Cleaning Composition. US patent 2197630; 1940

[34] Alexander P. Harry's Cosmeticology. The Principles and Practice of Modern Cosmetics. Vol. I. 6th ed. London: Leonard Hill Books; 1973

[35] Wenninger JA, Canterbery RC, McEwen GN Jr, editors. International Cosmetic Ingredient Dictionary and Handbook. Vol. 1-3. 8th ed. Washington, DC: Cosmetic, Toiletry, and Fragrance Association; 2000

[36] McDonald C, Richardson C. The effect of added salts on solubilization

by a nonionic surfactant. The Journal of Pharmacy and Pharmacology. 1981;**33**:38-39

[37] Kpogbemabou D, Lecomte-Nana G, Aimable A, Bienia M, Niknam V, Carrion C. Oil-in-water Pickering emulsions stabilized by phyllosilicates at high solid content. Colloids and Surfaces A: Physicochemical and Engineering Aspects. 2014;**463**:85-92

[38] Shalini S. Advantages and applications of nature excipients: A review. Asian Journal of Pharmaceutical Sciences. 2012;**2**(1):30-39

[39] Niazi SK. Handbook of Pharmaceutical Manufacturing Formulations: Compressed Solid Products. Cambridge: CRC Press; 2004

[40] King RE, Schwartz JB. Oral solid dosage forms. Remington's Pharmaceutical Sciences. 1985;**17**:1603-1632

[41] NICE. Drug reports (THR 15670/0020), (PL 14894/0297), (PL: 21727/0018-23). 2017. Available from: http://www.evidence.nhs.uk/ [Accessed: 12 September 2017]

[42] Aleanizy FS, Alqahtani F, Al Gohary O, El Tahir E, Al SR. Determination and characterization of metronidazole-kaolin interaction. Society of Professional Journalists. 2015;**23**:167-176

[43] Bonner WA, Flores J. On the asymmetric adsorption of phenylalanine enantiomers by kaolin. Currents in Modern Biology. 1973;**5**:103-113

[44] McElnay JC, D'Arcy PF, Throne O. Effect of antacid constituents, kaolin and calcium citrate on phenytoin absorption. International Journal of Pharmaceutics. 1980;**7**:83-88

[45] McElnay JC, Sidahmed AM, D'Arcy PF. Experimental modeling of drug absorption and drug absorption interactions. International Journal of Pharmaceutics. 1986;**31**:107-117

[46] Naggar VFB, Boraie NA, Shams-Eldeen MA. In vitro availability of promethazine–HCl in the presence of some commercial antacids. International Journal of Pharmaceutics. 1986;**28**:239-247

[47] Moustafa MA, Al-Shora HI, Gaber M, Gouda MW. Decreased bioavailability of quinidine sulphate due to interactions with adsorbent antacids and antidiarrheal mixtures. International Journal of Pharmaceutics. 1987;**34**:207-211

[48] Moustafa MA, Babhair SA, Kouta HI. Decreased bioavailability of some antipsychotic phenothiazines due to interactions with adsorbent antacid and antidiarrheal mixtures. International Journal of Pharmaceutics. 1987;**36**:185-189

[49] Moustafa MA, Gouda MW, Tariq M. Decreased bioavailability of propranolol due to interactions with adsorbent antacids and antidiarrheal mixtures. International Journal of Pharmaceutics. 1986;**30**:225-228

[50] Yu C, Bi E. Roles of functional groups of naproxen in its sorption to kaolinite. Chemosphere. 2015;**138**:335-339

[51] Gouda MW. Effect of an antidiarrheal mixture on the bioavailability of tetracycline and ampicillin. Abstracts of 21st Meeting of Academy of Pharmaceutical Sciences. 1976;**6**:117

[52] Albert KS, DeSante KA, Welch RD, DiSanto AR. Pharmacokinetic evaluation of a drug interaction between kaolin-pectin and clindamycine.

Journal of Pharmaceutical Sciences. 1978b;**67**:1579-1582

[53] Khalil SAH, Mortada LM, El-Khawas M. Decreased bioavailability of ampicillin and amoxycillin in presence of kaolin. International Journal of Pharmaceutics. 1984;**19**:233-238

[54] Khalil SAH, Mortada LM, Shams-Eldeen MA, El-Khawas MM. The In-vitro uptake of a low dose drug (riboflavin) by some adsorbents. Drug Development and Industrial Pharmacy. 1987;**13**(3):547-563

[55] Onyishi VI, Chime SA, Adibe CV. Formulation of pyridoxine hydrochloride sustained release capsules: Effect of propylene glycol co-solvent on the in vitro release. African Journal of Pharmacy and Pharmacology. 2013;7(15):809-815

[56] Wai KN, DeKay HG, Banker GS. Stability of vitamins A, B1, and C in selected vehicle matrices. Journal of Pharmaceutical Sciences. 1962;**51**(11):1076-1080

[57] Nokhodchi A. An overview of the effect of moisture on compaction and compression. Pharmaceutical Technology. 2005;**2**:46-66

[58] Ghebre-Sellassie I, Gordon RH, Middleton DL, Nesbitt RU, Fawzi MB. A unique application and characterization of Eudragit® E30D film coatings in sustained release formulations. International Journal of Pharmaceutics. 1986;**31**:43-54

[59] Nesbitt RU. Effect of formulation components on drug release from multiparticulates. Drug Development and Industrial Pharmacy. 1994;**20**:3207-3236

[60] Patel H, Stalcup A, Dansereau R, Sakr A. The effect of excipients on the stability of levothroxine sodium pentahydrate tablets. International Journal of Pharmaceutics. 2003;**264**:35-43

[61] Mohanachandran PS, Sindhumol PG, Kiran TS. Superdisintegrants an overview. International Journal of Pharmaceutical Sciences Review and Research. 2011;**6**(1):105-109

[62] Gopinath H, Venugopal KS, Shanmugasundaram S, Bada PK. A brief review on disintegrants. Journal of Chemical and Pharmaceutical Sciences. 2012;**5**(3):105-112

[63] Ward JB, Trachtenberg A. Evaluation of tablet disintegrants. Drug and Cosmetic Industry. 1962;**91**:35-36

[64] Lowenthal W. Disintegration of tablets. Journal of Pharmaceutical Sciences. 1972;**61**(11):1695-1711

[65] Kristensen J, Schæfer T, Kleinebudde P. Development of fast-disintegrating pellets in a rotary processor. Drug Development and Industrial Pharmacy. 2002;**28**(10):1201-1212

[66] Goyanes A, Souto C, Martínez-Pacheco R. Chitosan–kaolin coprecipitate as disintegrant in microcrystalline cellulose-based pellets elaborated by extrusions pheronization. Pharmaceutical Development and Technology. 2013;**18**(1):137-145

[67] Manivannan R, Parthiban KG, Sandeep G, Balasubramaniam A, Senthilkumar N. Multiparticlate drug delivery systems: Pellet & pelletization technique. Drug Discovery Today. 2010;**2**(5):233-237

[68] Srivastava S, Mishra G. Fluid bed technology: Overview and parameters for process selection. International Journal of Pharmaceutical Sciences and Drug Research. 2010;**2**(4):236-246

[69] Law MFL, Deasy PB. Effect of common classes of excipients on extrusion spheronization. Journal of Microencapsulation. 1997;**14**(5):647-657

[70] Deasy PB, Gouldson MP. In vitro evaluation of pellets containing enteric coprecipitates of nifedipine formed by non-aqueous spheronization. International Journal of Pharmaceutics. 1996;**132**:131-141

[71] Agrawal R, Naveen Y. Pharmaceutical processing – A review on wet granulation technology. International Journal of Pharmaceutical Frontier Research. 2011;**1**(1):65-83

[72] Shanmugam S. Granulation techniques and technologies: Recent progresses. BioImpacts: BI. 2015;**5**(1):55-63

[73] Chow AHL, Leung MWM. A study of the mechanisms of wet spherical agglomeration of pharmaceutical powders. Drug Development and Industrial Pharmacy. 1996;**22**(4):357-371

[74] Mallick S, Pattnaik S, Swain K. Current perspectives of solubilization: Potential for improved bioavailability. Drug Development and Industrial Pharmacy. 2007;**33**:865-873

[75] Panchagnula R, Bhardwaj V. Effect of amorphous content on dissolution characteristics of rifampicin. Drug Development and Industrial Pharmacy. 2008;**34**:642-649

[76] Qi S, McAuley WJ, Yang Z, Tipduangta P. Physical stabilization of low-molecular weight amorphous drugs in the solid state: A material science approach. Therapeutic Delivery. 2014;**5**(7):817-841

[77] Carretero MI, Pozo M. Clay and non-clay minerals in the pharmaceutical industry. Part I. Excipients and medical applications. Applied Clay Science. 2009;**46**:73-80

[78] Wedderburn DL. Baby preparations. In: Harry RG, editor. Harry's Cosmeticology. The Principles and Practice of Modern Cosmetics. Vol. I. 6th ed. London: Leonard Hill Books. 1973:543

[79] Felton LA, McGinity JW. Influence of insoluble excipients on film coating systems. Drug Development and Industrial Pharmacy. 2002;**28**(3):225-243

[80] Felton LA, Porter SC. An update on pharmaceutical film coating for drug delivery. Expert Opinion on Drug Delivery. 2013;**10**(4):421-435

[81] Zoglio MA, Maulding HV, Carstensen JT. Linearization of drug delivery from sustained-release dosage forms, synthetic gel systems. Drug Development and Industrial Pharmacy. 1996;**22**(5):431-437

[82] Carretero MI, Pozo M. Clay and non-clay minerals in the pharmaceutical and cosmetic industries. Part II. Active ingredients. Applied Clay Science. 2010;**47**:171-181

[83] Hermans MH. Wounds and ulcers: Back to the old nomenclature. Wounds. 2010;**22**(11):289-293

[84] Glick JB, Kaur RR, Siegel D. Achieving hemostasis in dermatology – Part II: Topical hemostatic agents. Indian Dermatology Online Journal. 2013;**4**(3):172-176

[85] Smith AH, Laird C, Porter K, Bloch M. Haemostatic dressings in prehospital care. Emergency Medicine Journal. 2013;**30**:784-789

[86] Pourshahrestani S, Zeimaran E, Djordjevic I, Kadri NA, Towler MR. Inorganic hemostats: The state-of-the-art and recent advances. Materials

Science and Engineering: C. 2016;**58**:1255-1268

[87] Z-Medica. Informative Website of QuikClot® Hemostatic Devices. 2017. Available from: http://www.quikclot.com/ [Accessed: 12 September 2017]

[88] Droy-Lefaix MT, Tateo F. Clays and clay minerals as drugs. In: Bergaya F, Theng BKG, Lagaly G, editors. Handbook of Clay Science. Amsterdam: Elsevier; 2006. pp. 743-752

[89] Leonard AJ, Droy-Lefaix MT, Allen A. Pepsin hydrolysis of the adherent mucus barrier and subsequent gastric mucosal damage in the rat: Effect of diosmectite and 16, 16 dimethyl prostaglandin E2. Gastroentérologie Clinique et Biologique. 1994;**8**:609-616

[90] Rozalen M, Huertas FJ, Brady PV. Experimental study of the effect of pH and temperature on the kinetics of montmorillonite dissolution. Geochimica et Cosmochimica Acta. 2009;**73**:3752-3766

[91] Habold C, Reichardt F, Le Maho Y, Angel F, Liewig N, Lignot J, et al. Clay ingestion enhances intestinal triacylglycerol hydrolysis and non-esterified fatty acid absorption. British Journal of Nutrition. 1985;**102**:249-257

[92] Kikouama OJR, Balde L. From edible clay to a clay-containing formulation for optimization of oral delivery of some trace elements: A review. International Journal of Food Sciences and Nutrition. 2010;**61**(8):803-822

[93] Constancio J, Pereira-Derderian DTB, Menani JV, De Luca Jr LA. Mineral intake independent from gastric irritation or pica by cell-dehydrated rats. Physiology & Behavior. 2011;**104**:659-665

[94] Jones BF, Galán E. Sepiolite and palygorskite. In: Bailey SW, editor. Hydrous Phyllosilicates. Reviews in Mineralogy. Vol. 19. Washington, DC: Mineralogical Society of America; 1988. pp. 631-674

[95] Christidis GE, Scott PW, Dunham AC. Acid activation and bleaching capacity of bentonites from the islands of Milos and Chios, Aegean, Greece. Applied Clay Science. 1997;**12**:329-347

[96] Vicente Rodríguez MA, López González JD, Bañares Muñoz MA. Acid activation of a Spanish sepiolite: Physicochemical characterization, free silica content and surface area of products obtained. Clay Minerals. 1994;**29**:361-367

[97] Aparicio P, Galán E. Mineralogical interference on kaolinite crystallinity index measurement. Clays and Clay Minerals. 1999;**47**:12-27

[98] Clark A, Ede R. Acute Diarrhoea: Causes and Recommended Management. 2011. Available from: www.prescriber.co.uk [Accessed: 12 September 2017]

[99] Primandini P, Hasanah AN, Adi WA, Budianto E, Sudirman S. The effect of calcination temperature on toxin adsorption materials for diarrheal diseases. Indonesian Journal of Materials Science. 2012;**13**(3):230-235

[100] Wardhana YW, Hasanah AN, Primandini P. Deformation and adsorption capacity of kaolin that is influenced by temperature variation on calcination. International Journal of Pharmacy and Pharmaceutical Sciences. 2014;**6**(3):1-2

[101] Pieszka M, Luszczynski J, Hedrzak M, Goncharova K, Pierzynowski SG. The efficacy of kaolin clay in reducing the duration and severity of heat' diarrhea in foals. Turkish Journal of Veterinary and Animal Sciences. 2016;**40**(3):323-328

[102] López-Galindo A, Viseras C. Pharmaceutical and cosmetic applications of clays. In: Wypych F, Satyanarayana KG, editors. Clay Surfaces: Fundamentals and Applications. Amsterdam: Elsevier Academic Press; 2004. pp. 267-289

[103] Medzhitov R. Origin and physiological roles of inflammation. Nature. 2008;**454**:428-435

[104] Charkoudian N. Mechanisms and modifiers of reflex induced cutaneous vasodilation and vasoconstriction in humans. Journal of Applied Physiology. 2010;**109**:1221-1228

[105] Cornejo-Garrido H, Nieto-Camacho A, Gómez-Vidales V, Ramírez-Apan MT, Angel P, Montoya JA, et al. The anti-inflammatory properties of halloysite. Applied Clay Science. 2012;**57**:10-16

[106] Caglar B. Structural characterization of kaolinite-nicotinamide intercalation composite. Journal of Molecular Structure. 2012;**1020**:48-55

[107] Cervini-Silva J, Camacho AN, Kaufhold S, Ufer K, Jesús ER. The anti-inflammatory activity of bentonites. Applied Clay Science. 2015;**118**:56-60

[108] Cervini-Silva J, Camacho AN, Palacios E, Angel P, Pentrak M, Pentrakova L, et al. Anti-inflammatory, antibacterial, and cytotoxic activity by natural matrices of nano-iron(hydr) oxide/halloysite. Applied Clay Science. 2016;**120**:101-110

[109] Awad ME, López-Galindo A, Setti M, El-Rahmany MM, Iborra CV. Kaolinite in pharmaceutics and biomedicine. International Journal of Pharmaceutics. 2017;**533**:34-48

[110] Tiwary AK, Poppenga RH, Puschner B. In vitro study of the effectiveness of three commercial adsorbents for binding oleander toxins. Clinical Toxicology. 2009;**47**:213-218

[111] Carraro A, De Giacomo A, Giannossi ML, Medici L, Muscarella M, Palazzo L, et al. Clay minerals as adsorbents of aflatoxin M1 from contaminated milk and effects on milk quality. Applied Clay Science. 2014;**88-89**:92-99

[112] Misyak SA, Burlaka AP, Golotiuk VV, Lukin SM, Kornienko PL. Antiradical, antimetastatic and antitumor activity of kaolin preparation Kremnevit. Galician Medical Journal. 2016;**23**(1):44-47

Chapter 7

Rheological Perspectives of Clay-Based Tailings in the Mining Industry

Ricardo I. Jeldres and Matías Jeldres

Abstract

The mining industry faces a significant problem in regions with water scarcity and has had to put in place new strategies to preserve its environmental and economic sustainability. An attractive option in recent years has been the direct use of seawater, avoiding the construction of reverse osmosis plants to desalinate. But, some operational complexities are the subject of discussion and research for engineers; for example, the difficulties by the high presence of complex gangues like clays and the location of the plants, far from the coast and at high altitude. The latter requires high investments in pumping, the only option in some cases. In this scenario, it is imperative to improve the efficiency of water use and advance to effective closures of water circuits. A critical stage is the thickening that allows water to be recovered from the tailings, reusing it in upstream operations. However, the performance of the tailings management is usually limited by the rheological properties of the thickened slurries, which impact on the discharge from the underflow of the thickeners, pumping energy costs, disposal on the tailings storage facilities (TSFs). This text describes the consequences caused by a saline medium on the rheological properties of clay-based tailings, analysing scenarios that allow tackling this operation.

Keywords: rheology, clays, seawater, thickening, tailings, water recovery

1. Introduction

On a global scale, mining is a relatively small consumer of water compared to the agricultural or forestry industry, however, it can generate a significant social and environmental impact, mainly in those companies located in arid regions as occurs in numerous operations in Chile, Australia, and South Africa [1, 2]. The industry has made great efforts to optimise the use of water where the proper tailings management is crucial towards advance to the effective closure of water circuits. Essential aspects are the quantity and quality of water recovered in thickeners, solid concentration of the thickened slurries, drainage capacity of the underflow in the lower cone of thickeners, the energy and water required for transport, and the disposal strategies in the tailings storages facilities (TSFs). The rheological properties largely determine the performance of all the aspects previously mentioned and in many cases, it defines the success that the design of a concentrator plant will have [3–5].

2. Tailings management

The different types of thickening technologies mostly establish the amount of water that can be extracted from the tailings, and hence the characteristics of the pulps that are transported to the TSFs. In the copper industry, conventional and high rate thickeners generate pulps between 50 and 60 wt% that have low yield stress (<40 Pa), while high-density and paste thickeners can make pulps over 65 wt% with yield stresses over 100 Pa [6, 7]. The thickened pulps are subsequently transported, usually by pumping to the tailings storage facilities (TSFs) [7].

The design of a tailings circuit should adequately consider the three stages involved (see **Figure 1**).

i. Performance in water recovery in solid-liquid separation stages (thickening mainly);

ii. Pumping of thickened tailings, whose costs fluctuate depending on the characteristics of both the pulps (rheological properties) and geography of the plants. In some instances the thickeners are located at a higher altitude than the TSFs, and gravity might assist the pumping. In contrast, on other cases the cost per transport may be decisive, especially when the distances are too long, the tailings have high density, or there are no gravitational advantages;

iii. Disposal methods and dewatering in TSFs, which primarily depend on the rheological behaviour of the slurries that consequently are a result of the thickening technologies.

Among the most advanced technologies to promote tailings dewatering are paste thickeners (**Figure 2(a)**). These equipment have a much higher lateral height than the other types of thickeners, a higher inclination of the discharge cone (30-45%), and the product is a tailing with a maximum concentration of solids. The pulps can

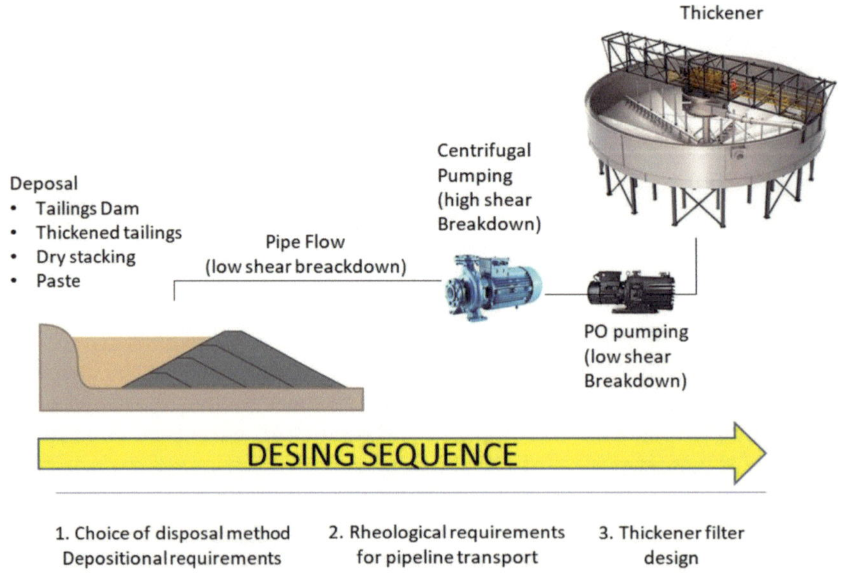

Figure 1.
Schematic representation of the stages involved in tailings management.

Figure 2.
(a) Paste thickener, (b) depositing of a tailing from a paste thickener.

reach solids concentrations near to 70 wt%, maximising the tailings dewatering, and facilitating their disposal in the TSFs (**Figure 2(b)**). Here comes the importance of the rheological properties of the slurries, since their pumping could incur excessive energy costs.

3. Rheological characterisation

The traditional method to characterise the rheological behaviour of tailings is through flow curves that are fitted to viscoplastic models such as Bingham or Herschel-Bulkley. Then, the yield stress is derived, which is widely used to describe, design, and control tailings processes in pipelines and beds. The accuracy of such measurements is controversial, and great care must be taken in some systems where precise equilibrium data is challenging to obtain (that is common for mining tailings). Nguyen and Boger [8] adopted the static yield stress vane measurement that is simple and has become widely used. Fisher et al. [9] later concluded that while the rough surfaces of the cup and bob geometries can be used successfully, infinite cup vane geometry alleviates all wall effects and it is a suitable method of determining yield stress and flow at steady-state behaviour of strongly aggregated particle suspensions.

Reograms are graphical representations of the response of the shear stress to variations in the angular strain rate, considering a material (suspension, in this case) between two parallel planes where one is moved, and the other remains immobile. The resolution of the Navier-Stokes equations, which describe the movement of the fluid, is simplified to an analytical form, taking care that the inertial forces are small compared to the viscous forces. Consequently, the viscometric flow is characterised by simple configurations in which the only relevant component of the stress tensor is the pure shear. Therefore the inverse problem for the viscosity can be solved directly by fitting the experimental measurements of some physical magnitude, as the torque M.

For the accurate description of the empirically obtained rheograms, it is necessary to consider the boundary conditions that are used to solve the Navier-Stokes equations in their simplified form. This means that particle sedimentation, the appearance of secondary flows (e.g. Taylor vortices) and phenomena such as wall slip should be avoided.

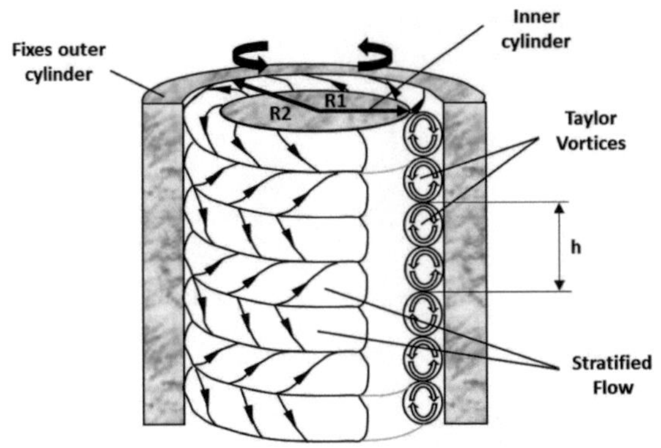

Figure 3.
Representation of Taylor vortices in a Couette geometry.

3.1 Taylor's vortices

Some low-viscosity slurries may involve a secondary flow driven by the inertia of the sample, forming a phenomenon called Taylor vortices (see **Figure 3**). When this happens, it is common for rheograms to observe a false increase in rheological properties or even shear thickening (dilatant) behaviours.

The appearance of the vortices is anticipated by the Taylor number (T), which in the case of concentric cylinders with internal cylinder rotation is described by the Eq. (1):

$$T = 2 \cdot \left(\frac{a_1}{a_2}\right)^2 \cdot \left(\frac{(a_2 - a_1)^4}{1 - \left(\frac{a_1}{a_2}\right)^2}\right) \cdot \left(\frac{\rho \cdot \omega}{\eta}\right)^2 \tag{1}$$

Where:

- a_1 inner radius
- a_2 outer radius
- ρ liquid density
- ω rotational speed of the inner cylinder in rad/s
- η fluid viscosity

3.2 Wall-slip

The obtaining of rheological parameters is based on the Navier-Stokes equation, whose expression is given by the Eq. (2):

$$\eta\left[\frac{\partial}{\partial r}\left(\frac{1}{r}\frac{\partial}{\partial r}(r v_\theta)\right)\right] = 0 \tag{2}$$

Obtaining an analytical solution to the expression requires a series of assumptions, for example, that the fluid moves in laminar flow, with streamlines moving in a radial direction, around the inner cylinder as shown in **Figure 3**. In particular, the fluid layer located in the contours is assumed to move at the same speed as the walls of the inner and outer cups. However, it is common for a wall slip phenomenon to occur, which is defined as the difference between the speeds of the wall relative to that of the fluid in the wall.

This phenomenon is caused by a thin layer of solution that forms on the walls and creates a wrong decrease in rheological properties. Obtaining the right rheological behaviour needs performing a rheogram correction, based on measurements made with geometries of different dimensions and shapes.

A method proposed by Yoshimura and Prud'homme [10], expects only two measurements, using geometries with different gaps but with the same ratio between the radii of the cup and the internal cylinder. According to this method, the slip velocity which corresponds to the stress τ^* from two angular velocity can be described as:

$$u_s(\tau^*) = \frac{\kappa}{\kappa+1}\left[\frac{\grave{U}_1 - \grave{U}_2}{\frac{1}{R_1} - \frac{1}{R_2}}\right] \quad (3)$$

Where:

- $u_s(\tau^*)$, slip velocity at the cup or bob surface
- κ, cup/bob ratio $= R_i / R_o$
- $\grave{U}$, angular velocity

Figure 4 shows the typical result of preparing measurements with different $\grave{U}$ ranges on each device. Yoshimura and Prud'homme [10] validated the proposed method with 1.9 wt% clay suspensions, measuring on different Couette devices. **Figure 4** also shows the experimental measurements and the corrected rheogram value.

Commonly, concentrated tailings show a non-Newtonian behaviour where they have an elastic limit. The yield stress, τ_y, is the critical shear stress that must be overcome before irreversible deformation and flow can occur. The yield stress is an engineering reality, although the rheology community debates hard about its

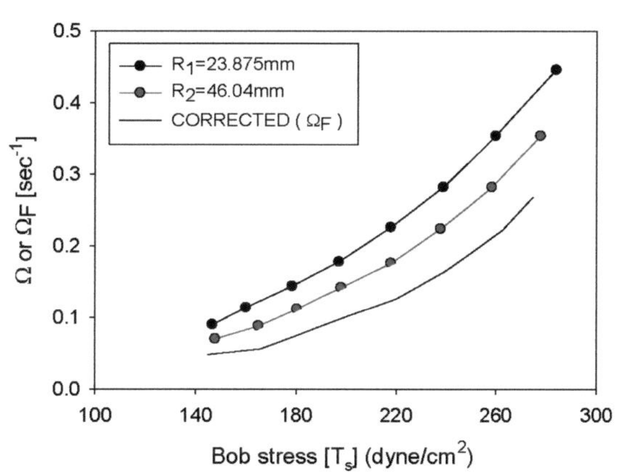

Figure 4.
Angular velocity (Ω) vs. bob stress for a 1.9 wt% clay suspension measured on different Couette devices. Also shown is the angular velocity corrected for wall slip (Ω_f) vs. bob stress (adapted from Yoshimura and Prud'homme [10]).

real existence [11, 12]. The truth is that its estimation is a routine practice to define control strategies for the management of mine tailings [13–15].

Yield stress depends on chemical interactions between particles, size distribution, and pulp density. This last aspect is critical in the search to maximise water recovery; however, the relationship between density and yield stress follows an exponential law where small changes in the percentage of solids lead to a significant rise in the value of yield stress. A schematic representation of the implications of thickening technology on the density and rheological behaviour of the pulps is shown in **Figure 5**. For example, when the underflow is thickened to too high concentration, the energy cost per pumping can reach prohibitive values, and operators may be required to dilute the thickened tailings, which means sacrificing water that could have been recirculated to upstream operations.

In the last time, some studies have integrated additional parameters in the discussion, especially analysing the viscoelastic behaviour of the pulps. Although to date there are no publications that directly use of viscoelastic parameters for the design of tailings circuits, their knowledge has allowed obtaining more information on the strength of the particle networks that make up the mineral slurries [16–18]. The behaviour of the sample is described through a viscous component, represented by the storage modulus (G'), and an elastic part, represented by the loss modulus (G'') [19]. The viscoelastic modulus can be obtained using dynamic oscillatory rheology methods, which are carried out by subjecting the sample to an oscillatory deformation $\gamma(t)=\gamma_{o}\sin(\omega t)$, obtaining the resulting stress as a function of time $\tau(t)=\gamma_{o}\left(G'(\omega)\sin(\omega t)+G''(\omega)\cos(\omega t)\right)$. G' is a measure of the material's stored energy and is therefore related to molecular events of an elastic nature, while G'' indicates the energy dissipated as heat, associated with viscous molecular events.

Lin et al. [18] analysed the structural changes of kaolinite particle junctions concerning the pH and concentration of solids in the pulp. Using small-amplitude oscillatory shear (SAOS) tests, the authors concluded that the gelation by attractive interactions in low-mass fractions changes to gelation by face-to-face interactions by increasing the particle mass fraction. Gelation by face-to-face interactions is stabilised by the repulsive electrostatic force between the faces of the disk-shaped particles. Based on a DLVO theory that is based on attractive forces between

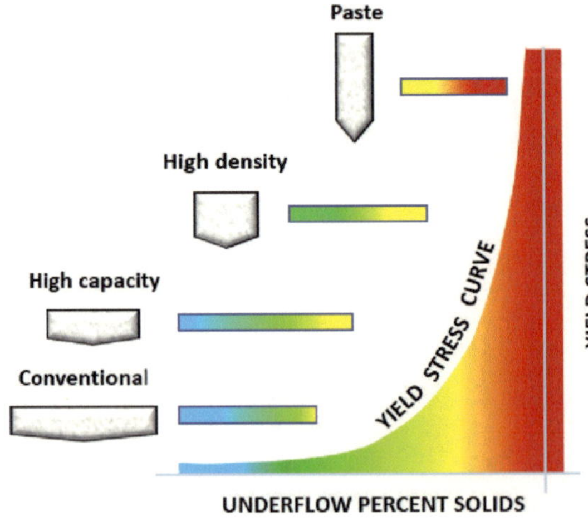

Figure 5.
Relationship between the type of thickener and properties of the underflows (solid concentration and yield stress).

particles, a modified model has been developed based on the repulsive electrostatic force between the faces of the platelets.

4. Rheological behaviour in saline environments

The current tailings dewatering challenges within the mining industry are the upstream use of low-quality water (like seawater) in processing and complex gangues like clays [20–22]. In the first case, the behaviour of the pulps in a highly saline environment must be faced, which is complex from a scientific point of view, since salinity significantly alters the interactions between the particle's surfaces, bringing important consequences in the rheological properties. They are strongly related to salinity and the type of salt [23]. For example, Reyes et al. [24] studied the rheological behaviour of magnetite tailings without flocculation, using mixtures of freshwater with seawater in different proportions. The authors were able to explain their findings in terms of electrostatic interactions, since they related the value of yield stress with the magnitude of the zeta potential, following the recommendation proposed two decades ago by Johnson et al. [25]. However, it should be noted that in complex saline systems, like seawater, there are many ions of different nature interacting simultaneously. In this case, the divalent cation speciation and the maker/breaker type may have an important role. For example, the formation of solid magnesium complexes at high pH ($pH < 10.5$) can cause the rheological properties of a quartz suspension to be reduced [26].

Jeldres et al. [27] analysed the viscoelastic behaviour of quartz suspensions prepared in monovalent brines. As seen in **Figure 6(a)**, there is a direct relationship between the size of the cations and the yield stress. The authors explained that silica has a more significant trend to agglomerate in the presence of larger ions like K^+, forming stronger particle networks compared to smaller salts like Na^+ and Li^+. According to fundamental studies, the surface of the silica would present a breaker-like behaviour [28], having a higher affinity with species of the same nature as is observed in **Figure 6(b)**, where the amount of adsorbed cation increases for larger sizes, following the order $Li^+ < Na^+ < Cs^+$. In the case of clays, the surface would have a higher affinity for water molecules, exhibiting a maker-like behaviour [29].

Clays are a consistent focus of research because they negatively impact almost all processes in the mining sector, including leaching, flotation, pulp transport, thickening, etc. [16, 30, 31]. These phyllosilicates can be classified according to their swelling (e.g. sodium montmorillonite) and non-swelling (e.g. kaolinite) character. This classification is made according to the response of the particles when they come into contact with water, being able to preserve their structure or increase their

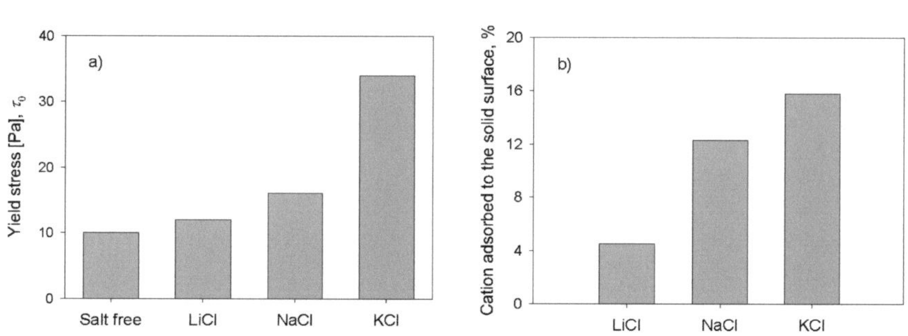

Figure 6.
Effect of the type of salt on (a) yield stress of quartz suspensions; (b) adsorption of the monovalent cation on the quartz surface (adapted from Jeldres et al. [27]).

apparent volume in solution. Kaolinite ($Al_2Si_2O_5(OH)_4$) (**Figure 7**) is composed of an octahedral sheet of aluminium hydroxide and a tetrahedral sheet of silica which are joined to form a basic 1:1 repeating unit. This clay has two crystallographically different surfaces: the faces that are negatively charged, and the edges, which vary their charge (anionic or cationic) depending on the pH that arises from the protonation or deprotonation of the aluminium (Al–OH) and silanol groups (Si–OH) in the exposed planes with hydroxyl termination. Due to their anisotropic structure and charge properties, clay sheets can form different types of interactions: face-to-face (FF), edge-to-edge (EE) and edge-to-face (EF).

Montmorillonite (**Figure 8**) is composed of an octahedral alumina sheet and two tetrahedral silica sheets that join to form a basic unit of repeating layers in a 2:1 ratio [32]. The central layer contains octahedral coordinated Al and Mg in the form of oxides and hydroxides and is surrounded by two external layers formed by tetrahedral coordinated silicon oxides. This clay has high chemical stability, a large surface

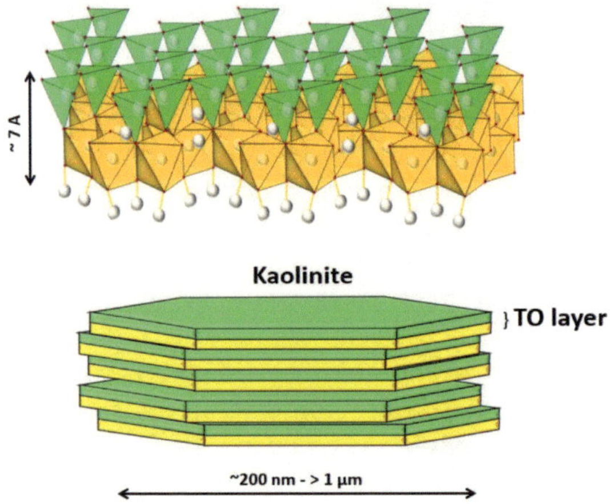

Figure 7.
Representation of the kaolinite structure. The green sheet indicates the tetrahedral layer of silicon (T) and the yellow sheet corresponds to the octahedral layer of alumina (O).

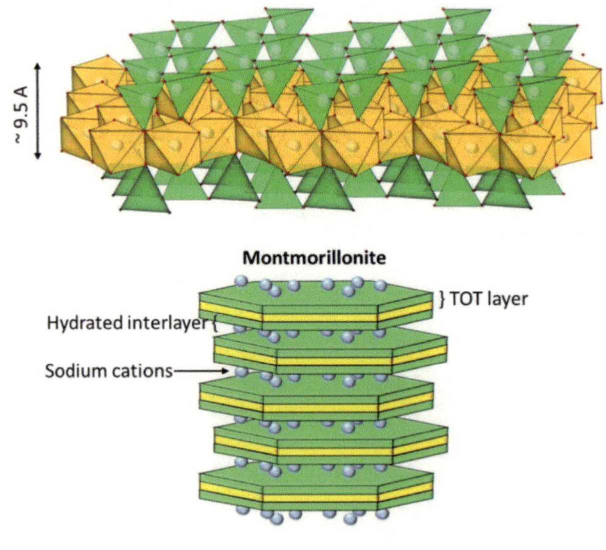

Figure 8.
Representation of the sodium montmorillonite structure. The green sheet indicates the tetrahedral layer of silicon (T) and the yellow sheet corresponds to the octahedral layer of alumina (O).

area and a top ion exchange capacity as a result of its weakly bound octahedral sheets. Unlike kaolinite, montmorillonite has a high capacity to swell in the presence of freshwater as a result of water molecules entering the middle of the layers [33]. In freshwater, the swelling effect of montmorillonite is more significant than in saline medium. This is because the cations being in saline water helps to neutralise the negative charges of the interlayer layers of the clay, reducing its electrostatic repulsion and therefore its separation distance [34–36]. This prevents the entrance of water molecules into the clay, notably impacting its swelling [37]. The high level of particles' swelling in freshwater implies an increase in the apparent volume concentration, which generates a considerable increase in the rheological properties of the slurries [38]. However, this characteristic is strongly influenced by the chemical nature of the water and the presence of ions in solution [39].

The chemical differences in the surface cause significant changes in the rheological properties of the pulps. For this reason, clay suspensions are exposed to dramatic changes concerning the pH of the suspension, mainly due to changes in electrostatic forces on surfaces. At low pH, there are anionic (faces) and cationic (edges) zones that generate attractive strong bonds between the particles, which clump together to form a "house of cards" structure. However, increasing the pH intensifies a greater electrostatic repulsion, causing the particles to disperse and the suspension to be more stable. The rheological consequences regarding pH have been widely addressed in the literature. Clays of 1:1 structure such as kaolinite generally show a behaviour that is decreasing concerning pH (**Figure 9**). In contrast, swelling clays such as montmorillonite have demonstrated the marked presence of a maximum at pH around 4 (**Figure 10**).

The electrostatic attraction between the faces and edges is the primary mechanism that gives strength to the bonds between clay particles. For this reason, the presence of salt causes a reduction in electrical charges, resulting in lower viscosity and yield stress. **Figure 11** shows the impact of adding 0.1 M of electrolytes to the suspension. The rheogram curve shifts downwards when salt is present, however, the values strongly depends on the type of cation, where divalent cations generate a greater impact as they are more efficient in compressing the double electrical layer. The effect at alkaline conditions is different. The slurries in freshwater have low rheological parameters but the compression of the double-layer may favour the particles' aggregation, forming both hydrogen and cationic bonds. The result is that the rheological properties increase in a saline medium like seawater; therefore, the management of clay-based tailings may involve higher energy costs. In this sense, it

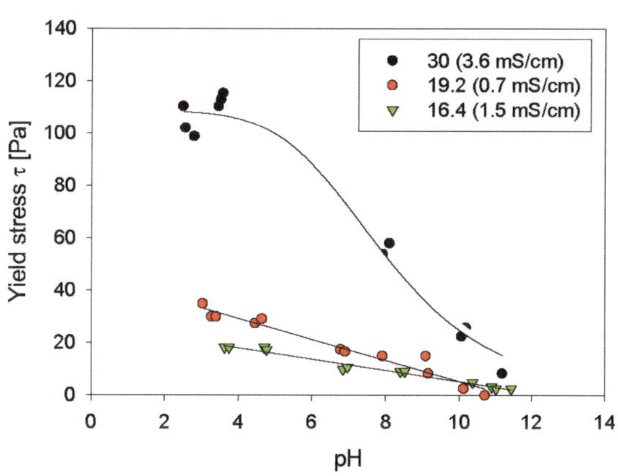

Figure 9.
Relationship between pH and yield stress for crown kaolin slurries at varied solid concentration (adapted from Au and Leong [40]).

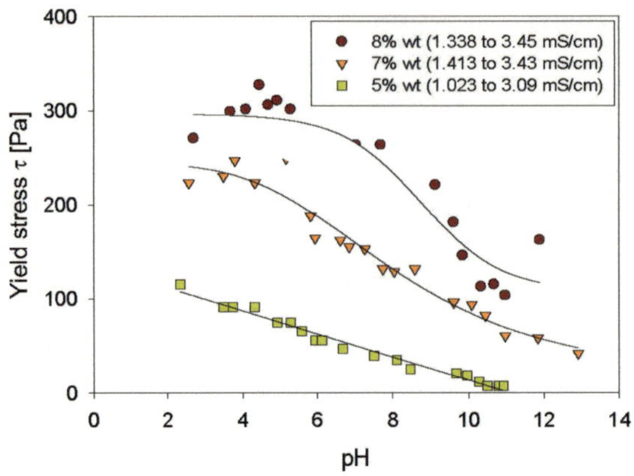

Figure 10.
Relationship between pH and yield stress for Na-montmorillonite slurries at varied solid concentration (adapted from Au and Leong [40]).

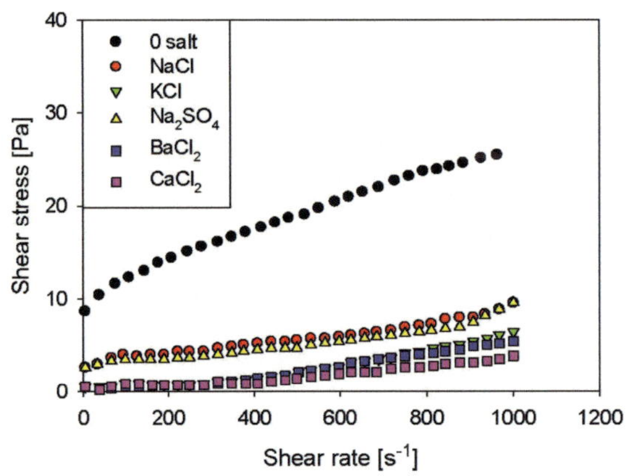

Figure 11.
Comparison between the effects of monovalent and divalent electrolytes on the flow curves of bentonite 1. Suspension: 8% bentonite in 0.1 M electrolyte solution (adapted from Abu-Jdayil [41]).

is necessary to find strategies to obtain pulps with rheological properties that simplify their transport. Contreras et al. [38] studied the effect of NaCl concentration on the rheological properties of synthetic tailings composed of mixtures of quartz and clays. Considering that the pulps were prepared at natural pH, it was found that salinity lowered the yield stress, as shown in **Figure 12**.

5. Chemical reagents

The permanent challenge for tailings management is to be able to manipulate their rheological properties, according to the plant bases. When seeking to facilitate the discharge of the tailings from the thickeners and to reduce the water and energy consumption involved in their transport by pumping, it is needed to find methods to reduce the values of yield stress and viscosity. Li et al. [42] used polycarboxylate copolymers synthesised in the treatment of kaolin suspensions. The results revealed the high capacity of the polymers to reduce the viscosity, due to the dispersion of the particles caused by steric and electrostatic effects, by increasing the anionic charge of the particles.

Du et al. [43] enlarged the electrostatic repulsion between bentonite particles by adding multiple charged phosphate-based reagents. The authors could reduce the yield stress to zero when it was a tetravalent or higher valence (see **Figure 13**). The results were explained by the changes in the electrostatic interactions, where the salts with higher valence were more pragmatic in reducing the negative zeta potential (**Table 1**).

However, when interactions occur in a highly saline medium, such as seawater, it is challenging to suggest strategies that drive an electrostatic repulsion since the high concentration of counter ions reduces the electrical double layer, making electrostatic changes less significant. For this reason, Robles et al. [44] recommended that the most efficient reagents are those that cause steric stabilisation of the

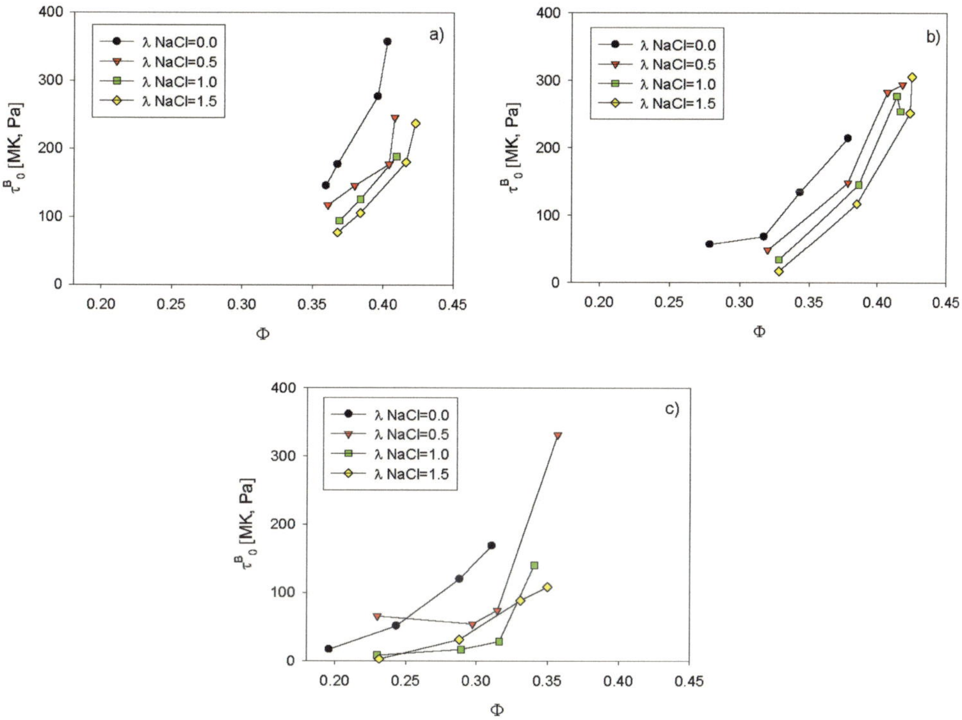

Figure 12.
Effect of the salinity on the yield stress of clay suspensions: (a) pure kaolin; (b) kaolin/Na-bentonite, 50/50; (c) pure Na-bentonite (adapted from Contreras et al. [38]).

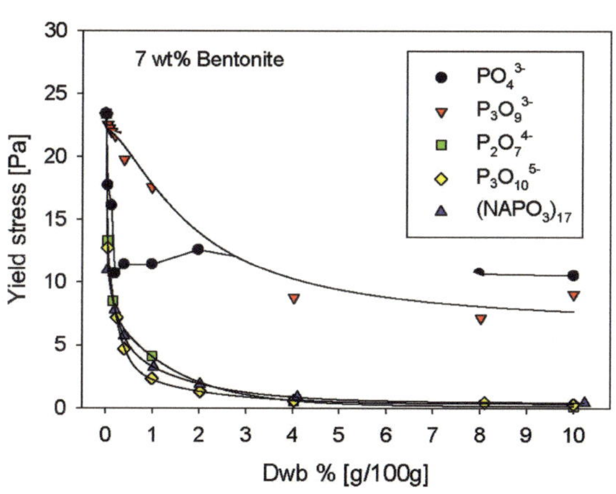

Figure 13.
Effect of phosphate-based additives on the yield stress of bentonite slurries (adapted from Du et al. [43]).

	None	PO_4^{3-}	$P_3O_9^{3-}$	$P_2O_7^{4-}$	$P_3O_{10}^{5-}$	$(NaPO_3)_{17}$
Z (mV)	-44	-48	-46.4	-64.4	-64.9	-66.1
pH	9.0	8.4	7.7	9.9	9.0	7.3

Table 1.
The effect of 10 dwb% phosphate-based additives on the zeta potential and pH of 7 wt% bentonite slurries (adapted from Du et al. [43]).

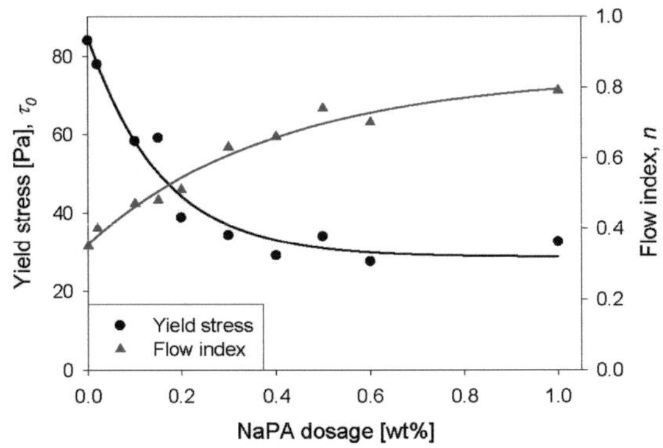

Figure 14.
Herschel-Bulkley parameters (yield stress and flow index) of kaolin pulp in seawater at pH 8 with varied sodium polyacrylate concentrations (Robles et al. [44]).

particles. The authors studied the influence of sodium polyacrylate of low molecular weight on the performance and viscoelasticity of kaolin pulps in seawater. It was shown that the reagent could reduce the strength of the bonds between the particles through steric stabilisation, considerably lowering the yield stress (**Figure 14**).

It was interesting that this polymer of low molecular weight provides promising results in a highly saline medium since the main reports in the literature have given its efficiency to the induction of higher anionic charges on the surfaces [45, 46].

6. Outlook

Tailings management in saline environments continues to be a challenging issue for plant design and operation, especially when significant clay content appears. Numerous factors enter the discussion, such as water quality, mineralogy (the type of clay), legal regulations, availability of water resources, etc. The need to improve the efficiency of this operation has increased in recent years, which has attracted greater scientific interest. Notable improvements have emerged from a technological point of view, in which the market offers increasingly robust equipment. That allows significant amounts of water to be extracted from the tailings and recycle them to upstream operations, with economic, environmental, and social benefits considering that many industries compete with neighbouring communities for disposing of the water resource. So reducing the water make-up means that more water would be available to the population.

Acknowledgements

The authors thank ANID/Fondecyt/11171036 and Centro CRHIAM Project ANID/FONDAP/15130015.

Conflict of interest

The authors declare no conflict of interest.

Author details

Ricardo I. Jeldres[1]* and Matías Jeldres[2]

1 Department of Chemical Engineering and Mineral Processing, University of Antofagasta, Antofagasta, Chile

2 Faculty of Engineering and Architecture, Arturo Prat University, Antofagasta, Chile

*Address all correspondence to: ricardo.jeldres@uantof.cl

References

[1] Northey, S. A.; Mudd, G. M.; Werner, T. T.; Jowitt, S. M.; Haque, N.; Yellishetty, M.; Weng, Z. The exposure of global base metal resources to water criticality, scarcity and climate change. Glob. Environ. Chang. 2017, 44, 109-124, doi:10.1016/j.gloenvcha.2017.04.004.

[2] Cisternas, L. A.; Gálvez, E. D. The use of seawater in mining. Miner. Process. Extr. Metall. Rev. 2018, 39, 18-33, doi:10.1080/08827508.2017.1389729.

[3] Boger, D. V. Rheology of slurries and environmental impacts in the mining industry. Annu. Rev. Chem. Biomol. Eng. 2013, 4, 239-257, doi:10.1146/annurev-chembioeng-061312-103347.

[4] Nguyen, Q. D.; Boger, D. V. Application of rheology to solving tailings disposal problems. Int. J. Miner. Process. 1998, 54, 217-233, doi:10.1016/S0301-7516(98)00011-8.

[5] Boger, D. V. Rheology and the Minerals Industry. Miner. Process. Extr. Metall. Rev. 2000, 20, 1-25, doi:10.1080/08827509908962460.

[6] Watson, A. H.; Corser, P. G.; Garces-Pardo, E. E.; López-Christian, T. E.; Vandekeybus, J. A comparison of alternative tailings disposal methods — the promises and realities. In Mine Waste 2010 -A.B. Fourie and R.J. Jewell (eds); Perth, 2010; pp. 499-514.

[7] Wang, C.; Harbottle, D.; Liu, Q.; Xu, Z. Current state of fine mineral tailings treatment: A critical review on theory and practice. Miner. Eng. 2014, 58, 113-131, doi:10.1016/j.mineng.2014.01.018.

[8] Dzuy, N. Q.; Boger, D. V. Direct yield stress measurement with the vane method. J. Rheol. (N. Y. N. Y). 1985, 29, 335-347, doi:10.1122/1.549794.

[9] Fisher, D. T.; Clayton, S. A.; Boger, D. V.; Scales, P. J. The bucket rheometer for shear stress-shear rate measurement of industrial suspensions. J. Rheol. (N. Y. N. Y). 2007, 51, 821-831, doi:10.1122/1.2750657.

[10] Yoshimura, A.; Prud'homme, R. K. Wall slip corrections for couette and parallel disk viscometers. J. Rheol. (N. Y. N. Y). 1988, 32, 53-67, doi:10.1122/1.549963.

[11] Hartnett, J. P.; Hu, R. Y. Z. Technical note: The yield stress—An engineering reality. J. Rheol. (N. Y. N. Y). 1989, 33, 671-679, doi:10.1122/1.550006.

[12] Barnes, H. A.; Walters, K. The yield stress myth? Rheol. Acta 1985, 24, 323-326, doi:10.1007/BF01333960.

[13] de Kretser, R.; Scales, P. J.; Boger, D. V. Improving clay-based tailings disposal: Case study on coal tailings. AIChE J. 1997, 43, 1894-1903, doi:10.1002/aic.690430724.

[14] Sofrá, F.; Boger, D. V Environmental rheology for waste minimisation in the minerals industry. Chem. Eng. J. 2002, 86, 319-330, doi:10.1016/S1385-8947(01)00225-X.

[15] Adiansyah, J. S.; Rosano, M.; Vink, S.; Keir, G. A framework for a sustainable approach to mine tailings management: disposal strategies. J. Clean. Prod. 2015, 108, 1050-1062, doi:10.1016/j.jclepro.2015.07.139.

[16] McFarlane, A. J.; Bremmell, K. E.; Addai-Mensah, J. Optimising the dewatering behaviour of clay tailings through interfacial chemistry, orthokinetic flocculation and controlled shear. Powder Technol. 2005, 160, 27-34, doi:10.1016/j.powtec.2005.04.046.

[17] Cruz, N.; Forster, J.; Bobicki, E. R. Slurry rheology in mineral processing unit operations: A critical review. Can.

J. Chem. Eng. 2019, 97, 2102-2120, doi:10.1002/cjce.23476.

[18] Lin, Y.; Phan-Thien, N.; Lee, J. B. P.; Khoo, B. C. Concentration dependence of yield stress and dynamic moduli of kaolinite suspensions. Langmuir 2015, 31, 4791-4797, doi:10.1021/acs.langmuir.5b00536.

[19] Barnes, H.A., Hutton, J.F., Walters, K. An introduction to rheology. Rheology Series (NL). Vol. 3.; Elsevier Science, Ed.; 1989; ISBN 9781493302611.

[20] Hernández, P.; Taboada, M.; Herreros, O.; Graber, T.; Ghorbani, Y. Leaching of chalcopyrite in acidified nitrate using seawater-based media. Minerals 2018, 8, 238, doi:10.3390/min8060238.

[21] Jeldres, R. I.; Uribe, L.; Cisternas, L. A.; Gutierrez, L.; Leiva, W. H.; Valenzuela, J. The effect of clay minerals on the process of flotation of copper ores - A critical review. Appl. Clay Sci. 2019, 170, 57-69, doi:10.1016/j.clay.2019.01.013.

[22] Liu, D.; Edraki, M.; Fawell, P.; Berry, L. Improved water recovery: A review of clay-rich tailings and saline water interactions. Powder Technol. 2020, 364, 604-621, doi:10.1016/j.powtec.2020.01.039.

[23] Jeldres, R. I.; Piceros, E. C.; Wong, L.; Leiva, W. H.; Herrera, N.; Toledo, P. G. Dynamic moduli of flocculated kaolinite sediments: effect of salinity, flocculant dose, and settling time. Colloid Polym. Sci. 2018, 296, 1935-1943, doi:10.1007/s00396-018-4420-x.

[24] Reyes, C.; Álvarez, M.; Ihle, C. F.; Contreras, M.; Kracht, W. The influence of seawater on magnetite tailing rheology. Miner. Eng. 2019, 131, 363-369, doi:10.1016/j.mineng.2018.11.037.

[25] Johnson, S. B.; Franks, G. V.; Scales, P. J.; Boger, D. V.; Healy, T. W. Surface chemistry–rheology relationships in concentrated mineral suspensions. Int. J. Miner. Process. 2000, 58, 267-304, doi:10.1016/S0301-7516(99)00041-1.

[26] Jeldres, M.; Piceros, E.; Robles, P. A.; Toro, N.; Jeldres, R. I. Viscoelasticity of quartz and kaolin slurries in seawater: Importance of magnesium precipitates. Metals (Basel). 2019, 9, 1120, doi:10.3390/met9101120.

[27] Jeldres, R. I.; Piceros, E. C.; Leiva, W. H.; Toledo, P. G.; Quezada, G. R.; Robles, P. A.; Valenzuela, J. Analysis of silica pulp viscoelasticity in saline media: The effect of cation size. Minerals 2019, 9, 1-15, doi:10.3390/min9040216.

[28] Quezada, G. R.; Rozas, R. E.; Toledo, P. G. Molecular dynamics simulations of quartz (101)-water and corundum (001)-water interfaces: Effect of surface charge and ions on cation adsorption, water orientation, and surface charge reversal. J. Phys. Chem. C 2017, 121, doi:10.1021/acs.jpcc.7b08836.

[29] Quezada, G. R.; Rozas, R. E.; Toledo, P. G. Ab initio calculations of partial charges at kaolinite edge sites and molecular dynamics simulations of cation adsorption in saline solutions at and above the pH of zero charge. J. Phys. Chem. C 2019, 123, 22971-22980, doi:10.1021/acs.jpcc.9b05339.

[30] Castillo, C.; Ihle, C. F.; Jeldres, R. I. Chemometric optimisation of a copper sulphide tailings flocculation process in the presence of clays. Minerals 2019, 9, 582, doi:10.3390/min9100582.

[31] Ma, X.; Fan, Y.; Dong, X.; Chen, R.; Li, H.; Sun, D.; Yao, S. Impact of clay minerals on the dewatering of coal slurry: An experimental and molecular-simulation Study. Minerals 2018, 8, 400, doi:10.3390/min8090400.

[32] Bergaya, F.; Theng, B.; Lagaly, G. Handbook of clay science; 2006;

[33] Kozak, M.; Domka, L. Adsorption of the quaternary ammonium salts on montmorillonite. J. Phys. Chem. Solids 2004, 65, 441-445, doi:10.1016/j.jpcs.2003.09.015.

[34] Gorakhki, M. H.; Bareither, C. A. Salinity effects on sedimentation behavior of kaolin, bentonite, and soda ash mine tailings. Appl. Clay Sci. 2015, 114, 593-602, doi:10.1016/j.clay.2015.07.018.

[35] Zhou, Z. Swelling clays in hydrocarbon reservoirs: The bad, the less bad, and the useful; UNITAR International Conference on Heavy Crude and tar Sands; China National Petroleum Corporation: Petroleum Industry Press, 1998;

[36] Zhang, J. F.; Zhang, Q. H.; Maa, J. P. Y. Coagulation processes of kaolinite and montmorillonite in calm, saline water. Estuar. Coast. Shelf Sci. 2018, 202, 18-29, doi:10.1016/j.ecss.2017.12.002.

[37] Elmashad, M. E.; Ata, A. A. Effect of seawater on consistency, infiltration rate and swelling characteristics of montmorillonite clay. HBRC J. 2015, 12, 175-180, doi:10.1016/j.hbrcj.2014.12.004.

[38] Contreras, S.; Castillo, C.; Olivera-Nappa, Á.; Townley, B.; Ihle, C. F. A new statistically-based methodology for variability assessment of rheological parameters in mineral processing. Miner. Eng. 2020, 156, 106494, doi:10.1016/j.mineng.2020.106494.

[39] Dong, Y.; Li, H.; Fan, Y.; Ma, X.; Sun, D.; Wang, Y.; Gao, Z.; Dong, X. Tunable dewatering behavior of montmorillonite suspension by adjusting solution pH and electrolyte concentration. Minerals 2020, 10, 293, doi:10.3390/min10040293.

[40] Au, P.-I.; Leong, Y.-K. Surface chemistry and rheology of slurries of kaolinite and montmorillonite from different sources. KONA Powder Part. J. 2016, 33, 17-32, doi:10.14356/kona.2016007.

[41] Abu-Jdayil, B. Rheology of sodium and calcium bentonite–water dispersions: Effect of electrolytes and aging time. Int. J. Miner. Process. 2011, 98, 208-213, doi:10.1016/j.minpro.2011.01.001.

[42] Moreno, P. A.; Aral, H.; Cuevas, J.; Monardes, A.; Adaro, M.; Norgate, T.; Bruckard, W. The use of seawater as process water at Las Luces copper-molybdenum beneficiation plant in Taltal (Chile). Miner. Eng. 2011, 24, 852-858, doi:10.1016/j.mineng.2011.03.009.

[43] Du, M.; Liu, J.; Clode, P.; Leong, Y.-K. Microstructure and rheology of bentonite slurries containing multiple-charge phosphate-based additives. Appl. Clay Sci. 2019, 169, 120-128, doi:10.1016/j.clay.2018.12.023.

[44] Robles, P.; Piceros, E.; Leiva, W. H.; Valenzuela, J.; Toro, N.; Jeldres, R. I. Analysis of sodium polyacrylate as a rheological modifier for kaolin suspensions in seawater. Appl. Clay Sci. 2019, 183, 105328, doi:10.1016/j.clay.2019.105328.

[45] Labanda, J.; Llorens, J. Influence of sodium polyacrylate on the rheology of aqueous Laponite dispersions. J. Colloid Interface Sci. 2005, 289, 86-93, doi:10.1016/j.jcis.2005.03.055.

[46] Huang, G.; Pan, Z.; Wang, Y. Synthesis of sodium polyacrylate copolymers as water-based dispersants for ultrafine grinding of praseodymium zirconium silicate. Colloids Surfaces A Physicochem. Eng. Asp. 2018, 558, 591-599, doi:10.1016/j.colsurfa.2018.08.027.